DARWIN'S DEVICES

DARWIN'S DEVICES

What **EVOLVING ROBOTS**
Can Teach Us About the History of Life
and the Future of Technology

JOHN LONG

BASIC BOOKS

A Member of the Perseus Books Group
New York

Published by Basic Books,
A Member of the Perseus Books Group

Books published by Basic Books are available at special discounts for bulk purchases in the
United States by corporations, institutions, and other organizations. For more information,
please contact the Special Markets Department at the Perseus Books Group, 2300 Chestnut
Street, Suite 200, Philadelphia, PA 19103, or call (800) 810-4145, ext. 5000, or e-mail
special.markets@perseusbooks.com.

Designed by Jeff Williams

Library of Congress Cataloging-in-Publication Data
Long, John, 1964 Jan. 12–
    Darwin's devices : what evolving robots can teach us about the history of life and the
future of technology / John Long.
        p. cm.
    Includes bibliographical references and index.
    ISBN 978-0-465-02141-3 (hardback)—ISBN 978-0-465-02928-0 (e-book)
    1. Evolutionary robotics. 2. Evolution (Biology)—Simulation methods. 3. Technology—
Forecasting.   I. Title.

TJ211.37.L66 2012
629.8'92—dc23
                                                                                2011051804

10 9 8 7 6 5 4 3 2 1

To Mom, Dad, Ann, Sam,
Marian, Tamasin, and Madeleine

CONTENTS

chapter 1

WHY ROBOTS?

I AM A BIOLOGIST, AND I STUDY ROBOTS. BUT AS SOON AS I started describing my research to other people, it was clear I was in trouble. I was speaking to a longtime friend and colleague about a biology grant I'd just gotten from the National Science Foundation to build robots when he stopped me in my tracks. "What do robots have to do with biology?" he asked. I knew then, with the certainty that only dread can provide, that this was an inescapable question—it was the issue that would come up first from now on, every time one of my students or I presented our strange new work to biologists.

What's the problem? First and foremost, biologists do not study robots. They work on organisms—living things, their environments, and their evolutionary history. They use machines as tools to ascend a rainforest canopy, as instruments to measure biomechanical properties, as modes of transportation to collect fish from a coral reef. As my friend had stated so succinctly, machines in general, and robots in particular, have nothing to do with biology—from his point of view. Not from mine.

I tried to fight back, blurting out the well-rehearsed line that I'd included in the grant proposal: "We use robots to model extinct vertebrates." With that being not so much an answer as a statement of intent, I got a raised right eyebrow and then a gentle "Well, I hope it works out for you." Communication over and out.

I needed a better answer.

No matter how cool robots are (to me), their swaggering presence wasn't enough to justify their usefulness in biology. And this was a problem not only for me—at least I had a grant—but also for the undergraduate researchers in my lab, who would rather avoid being recognized as having been trained by a known kook. So we talked it over until we hit upon a solution that has proven to work for about half the biologists we encounter.

We decided to equate different models of biological systems: those run on computers and those run on robots. Both are machines, after all, and computers are already used in almost every branch of biology, modeling—among myriad other things—neural networks, predator-prey interactions, virus evolution, and perambulating *Tyrannosauruses*. In fact, "computational biology" is the hot field right now, the bull's eye on the what-to-be-working-on-if-you-want-a-job-in-academia dartboard.

Robots—mobile ones, anyway—are essentially self-propelled computers. They are machines that run sets of instructions—their software—and produce an output. Certainly, the outputs seem different. Computers output binary bits that we use to represent numbers, and those numbers, in turn, represent everything from screen colors to mathematical formulae to electronic books. Robots output what we recognize as behavior, but underlying it all are the same bits.

That's not to say there might not be important distinctions between a robot and a computer. Jeff Staten, a senior engineer at IBM, says a robot "is a computer that's inside out." Jeff's point is that computers today are networked and make decisions based on input from other computers most of the time and humans tapping at a keyboard only some of the time. A robot, although it has a computer inside, makes decisions on its own, with information gathered only through its sensors. The messages the robot receives and sends are physical.

The mobile robot, unlike most computers, can be autonomous. What autonomous robots have, which computers don't, is agency.

Agency is what human observers ascribe to anything, organic or artificial, that appears from an external perspective to act on its own. For those of us working in the world of artificial intelligence and cognitive science, an agent can be an organism or a machine. Humans are agents. My dog, Kooka, is an agent. And so long as there is no unseen human pulling the strings of remote control, robots are agents too.

An autonomous robot is an agent using its own sensory inputs to perceive the world, make decisions about how to move, and, in turn, having those movements affect how it perceives the world. This constant feedback between what an agent perceives and how it moves is what my colleague Ken Livingston and I call a perception-action feedback loop. Multiple perception-action loops in an agent can be operating in parallel, working in combination, fusion, or competition. What we observe the agent doing—moving and interacting with its immediate environment—is what we define as behavior.

An agent's behavior is its computational output. "Behavior emerges from the agent-environment interaction," as Livingston is wont to say. And behavior, by this definition, is something that autonomous robots have that computers do not. Oops. Have we just defined ourselves into an identity error that invalidates the logic of our first response? No and yes. No, because autonomous robots have, as part of their agency, embedded computers. Yes, because robots are more than simply computers that move.

Autonomous agency is, ultimately, the answer to my colleague's question, what do robots have to do with biology? They enable us to build models of how organisms behave. Of course, building models raises another question.

It turns out that, much like most biologists think they shouldn't be studying machines, many think they shouldn't be studying models, either. One criticism of models or simulations of any kind, instantiated on either a computer or a robot, is that they are, at best, artificial systems that merely copycat the outward behavior of the biological system. Models, the argument goes, fail as true or accurate representations of the underlying causal phenomena because the underlying functional mechanisms are different from those operating in the system of interest. As Norbert Weiner, the founder of the field of cybernetics, is alleged to have said, "The best model of a cat is a cat." Although this is not quite like saying that to understand cats, you can only study cats, some people do jump to that conclusion.

With this cats-only criticism in mind, my colleagues and I jump to a different conclusion about models: you have to be very careful when you build them. You have to take care to explain what you are trying to do, how you intend to do it, and how you are discriminating between a bad model and a good one. Bad models won't be anything like cats, and therefore, they will perhaps tell you something about your ability to make a noncat—but little else. Good models, argues Barbara Webb, a biologist at the University of Edinburgh and one of the founders of the field of biorobotics, are those with explicitly defined goals and goals that are attained. Some models are meant to behave like the targeted system. In those cases a close or perfect behavioral match between your robot and its biological target—if it walks like a cat and meows like a cat—means that you have a good model.

As a real example, Sarah Partan, an animal behaviorist at Hampshire College, wanted to study how squirrels respond to the behavior of other squirrels, so she built a robot squirrel that flicks its tail and adjusts its posture. For Partan, a good squirrel model is able to trick the real squirrels into responding to the robot as if the robot were a squirrel. Happily for Partan, the trick worked.

Behaviorally speaking, Partan's robotic squirrel is a good model of body posture and tail motion. At the same time, it is obviously a bad model of the neuromuscular mechanisms involved in limb motion. However, to model neuromuscular mechanisms wasn't her intent. If it was, you'd judge the goodness of her neuromuscular mechanisms model not by how well it elicits tail flicks in other squirrels but rather by how closely it matches the underlying functional mechanisms used in muscles and nerves. If the artificial muscles worked like biological ones, generating peak force at only one length and for short periods of time, then she'd have modeled that mechanism accurately, irrespective of whether or not her robotic squirrel can dupe real squirrels.

Different models, then, serve different goals. Webb enumerates seven: emulating behavior and mechanism (both of which we've just seen) as well as abstractness, medium, generality, level, and, particularly important for us, whether or not the model tests a hypothesis—that is, an idea that you have about the biological system.

For Webb, if you are interested in any particular aspect of cats and you have learned as much from real cats as you can, then go ahead and model a cat. I think this would be Weiner's position as well. But if your goal is to learn more about cats, Webb says, avoid the temptation to build something cool just for the sake of building it. There must be a specific target: you shouldn't just try to build something cat-like. Her criticism is aimed at two fascinating fields, adaptive behavior and artificial life, in which many workers model invented animals, called animats. Although animats illuminate general operating and cognitive principles, Webb argues that the adaptive behavior and artificial life approaches do little to test specific hypotheses about how real animals work. For her, biologists building animats to test biological hypotheses usually walk away empty handed.

Webb's critique of adaptive behavior and artificial life gets to the heart of my colleague's skepticism. He intuited two of her points related to the value of any model to a biologist. First, your model must have a specific biological target. Second, your model must be relevant and must enable testing a hypothesis about the targeted system.

Of course, Webb's critique also raises some problems for our response to the skeptics—that we should be able to use robots because robots are essentially computers. If computer models aren't necessarily biologically relevant, then our robots may not be either. So our critique of the critiquer became this: he was asking the wrong question, or at least asking a sufficiently vague question to allow literalists like us to misunderstand. Instead of asking, "Why robots?" our skeptic ought to be asking, "What is the scientific purpose of your model, and why is your model in the form of a physically embodied robot?" But then there's probably a question you'd like to ask of me: how did a biologist ever find himself sticking up for robots? The answer's quite simple: it was for the love of fish.

FOR THE LOVE OF FISH

I've loved fish since I was a child, when I first saw Jacques Cousteau's underwater world on television. I followed the wake of swimming fish and other aquatic vertebrates through college, graduate school, and, now, into a career. And there was a lot to learn in the flesh. With Steve Wainwright of Duke University and Mark Westneat of the Field Museum, I've scuba dived to videotape the propulsive oscillation of two hundred–pound blue marlin. With Mark, Melina Hale of the University of Chicago, and Matt McHenry of the University of California, I've outfitted rainbow trout, bowfin, longnose gar, and African bichir with tiny instruments to measure muscle function during escape maneuvers. With Wyatt Korff of the Janelia Farm Research Campus, I've used high-speed video to investigate how Wyatt's trained Amazonian arawana can propel themselves out of the water to catch food in midair. With Lena Koob-Emunds and Tom Koob, I've worked at the Mount Desert Biological Laboratory to study the biomechanics of slimy pink hagfish so that we can learn about swimming without a vertebral column. Finally, with Marianne Porter of the University of California, I've measured the mechanical properties of the backbones of dogfish sharks to see how skeletons transmit force.

I love real fish, and more than twenty years' work has shown me much about their shape and structure, how they move, and how they evolved. But real fish only reveal some of their secrets. As with any science, we are limited by what we can and cannot observe and measure. Sometimes we lack an instrument or a technology. At other times we lack the right fish. Take, for example, the giant blue marlin. They die in captivity, so studying them in a lab was impossible. So we went blue-water diving not because we wanted to (it's very expensive and dangerous work) but because we had no other choice. We had to film blue marlin from a distance and leave some of our questions unanswered.

Well, not entirely. We could've studied marlin in other ways, but other questions would've been left unanswered. Say you want to know what's going on inside a blue marlin when it swims. You could, as Barbara Block of Stanford University did, build a team of engineers and physiologists to design tiny instruments that can be implanted quickly in a marlin that you've brought alongside a boat using hook and line. These instrument tags carry their own computer, power pack, and broadcast system, collecting data and sending signals back to a ship or satellite. Block can measure the marlin's body temperature, muscle activity, speed, and depth as it moves freely about its oceanic cabin.

But for all Block's approach reveals about physiology, the method reveals nothing about the biomechanics of marlin backbones—and that's what I wanted to study. The backbone, or vertebral column, runs from an animal's head to tail, and its presence is one of the signatures of the vertebrates, a group of animals to which amphibians, birds, fish, mammals, and reptiles—some fifty-eight thousand species—all belong. In a fish the backbone prevents the body from shortening while also allowing it to bend, and it gives the whole body important mechanical features, such as the ability to store and release energy elastically like a spring. My pursuit of this question is what would ultimately send me headlong into the world of artificial intelligence and robots.

FROM THE FIELD TO THE LAB

I first met Block—and the blue marlin—back in 1986, when she, as a newly minted PhD from Knut Schmidt-Neilsen's lab at Duke University, convinced me, a newbie PhD wannabe in Steve Wainwright's lab, to work on the biomechanics of marlin vertebral columns in the laboratory. Under the guise of buying me a cup of coffee at the Ninth Street Bakery, Block pulled me out of the lab my first day so she could expound the virtues of the marlin.

Of all the fish, she explained in the car, marlin are the best, the fastest, biggest, coolest predators in the sea. "Think tuna are fast?" she asked rhetorically as we pulled into the parking lot. "Well, marlin eat tuna!" As we walked across the street to the bakery, she went for the kill. "Have you seen the vertebral column of a marlin?" she asked, sounding like a minister in the First Church of Poseidon. I knew the proper response: "No, I have not seen the vertebral column of a marlin. What does it look like?"

She introduced me to the mysteries of the marlin's backbone. "It's not like a bunch of little bones linked together, like pearls on string, that you see in regular bony fish," she said. "The vertebral column of a marlin looks like a piece of wood, a long pine board, a one-by-six, with bones overlapping, bones welded together with collagenous connective tissue to form a single, giant spring." She paused for effect. "And this spring works to store and release energy, the energy that powers the high speeds and spectacular leaps of marlin."

I shuffled forward in the line, unable to muster words. Only images came to mind: marlin leaping and spinning above white caps, and terrified tuna, swimming for their lives but unable to avoid the explosive charges of the spring-loaded marlin. Block waited for a moment, paid for our coffee and muffins, and guided me to a table. Signaling with her hand for me to eat something, she gave me a chance to return from my reverie. Then she said, knowing the answer, "So. Are you in?" I gushed, "Absolutely!"

Giddiness gave way to the not harsh but practical realities of scientific research. Working in Wainwright's lab, I spent the next five years chasing after the elusive blue marlin, literally and figuratively. I wanted to measure their vertebral column's mechanical properties, features like stiffness—related to how much the vertebral column would resist the magnitude of bending and how much spring energy it would store—and energy loss—related to how much the vertebral column would resist the speed of bending and how much energy would be lost as heat. If stiffness is large compared to energy loss, the backbone would be a spring; if the stiffness is relatively small compared to energy loss, the backbone would work as a brake. If I could measure stiffness and energy loss of the vertebral column over a range of motions and speeds, I knew that I could have some idea of what Wainwright calls "mechanical design"—in this case how the mechanical properties of the vertebral column allow it to operate as the blue marlin swims or leaps.

Unable to buy an off-the-shelf marlin-testing machine (they don't exist), I had to design, build, and calibrate a customized vertebral column bender. My DIY guru for this challenge was Steven Vogel, also at Duke, who helped me brainstorm designs and taught me the difference between a DC brushless motor and a servo one. Once I had a working bending machine in place, Block and Wainwright helped get me and my machine out to the big island of Hawaii and the Pacific Gamefish Research Foundation.

On the Kona side of the island deepwater blue marlin are caught by recreational fishers literally in sight of the steep-sloped volcanic beaches, where, hat in hand, I would beg at the local fish houses for the castaway vertebral columns. Once I had one, I was unable to sleep until I had put each individual motion segment, consisting of two vertebrae and the intervening joint, through a series of mechanical tests. I'd bend the segments with varying frequency and amplitude, just like the marlin would have done as it hit the turbo button to pursue a tuna. To get a sense of what parts of the bone and joint structure helped cause changes in stiffness and energy loss along the column, I also measured the size and shape of each joint and the adjoining vertebrae. Lather, rinse, repeat. After several weeks I had tested the vertebral columns from six different marlin ranging in length from four to seven feet and weighing from thirty-six to more than two hundred pounds.

THE MECHANICAL DESIGN OF THE MARLIN'S BACKBONE

Back at Duke I began running the raw data through the Newtonian equations of motion that govern the relation between the bending motions the machine imposed on each joint and the bending torque each joint developed in resistance to that imposed motion. Looking over the range of joint positions, bending frequencies, and amplitude, I began seeing some very interesting patterns. The biggest surprise was that the tail, which looks from the video we took of marlin swimming to be the most flexible part of the body, actually has the stiffest part of the vertebral column. Talking with Wainwright, we realized that this was a counterintuitive result only because we were thinking of a jointed column as a series of bony blocks and frictionless hinges. If instead the joints—the hinges—were very stiff because of all of the overlapping bits of bone that Block had talked up, then the joints themselves appeared to be capable of storing energy as they bend.

But were those same joints able to release that spring energy as they unbent? This is where the energy loss came into play, and the marlin played a trick on us again. With simple ideas of springs in our heads, we had been thinking that as the marlin swam faster, increasing the frequency of their tail beats, their vertebral column would become even more spring-like, storing and releasing more elastic energy to match the power that the faster speeds demanded. We expected stiffness to increase and the energy loss to decrease. Just the opposite occurred.

To make sense of these surprises in the biological context of the swimming marlin, we put this information about mechanical properties into a mental, conceptual model of what we thought might be going on inside the marlin. Our guess was that as marlin increased swimming speed, the vertebral column would be adjusting its mechanical behavior, switching gradually from a spring to a spring with a brake. This spring-and-brake mechanism is exactly how the shock absorbers in your car work, with the spring resisting the initial bump, giving way gently, and then returning the wheel to its place on the road. At the same time, the brake, or what we call a dashpot in a shock absorber, uses fluid to dampen the spring's motion, keeping the spring from bouncing the car vertically after that first bump.

Sounds reasonable, doesn't it? We've only one problem: a backbone in my machine doesn't necessarily act like one in a dynamically operating animal. That problem is what drove us to Hawaii in search of underwater footage of swimming marlin. We wanted to see how living marlin moved their bodies, how fast they beat their tails, and how much they bend their backbones. Knowing this would let us make a good guess about how the vertebral column is operating during swimming, but it still wouldn't let us measure the backbone directly as it bent nor evaluate how the muscles responsible for driving the bending do their work. Nor, for that matter, would we be gauging the complex forces from the surrounding water interacting with the undulating body.

The blue marlin is the poster child for problems with the biomechanical approach. And it wasn't that we were stymied just because we couldn't bring the fish into the lab to implant measuring devices. Even if we could, problems would remain. For example, when we try to directly measure the forces that bend the vertebral column of living, swimming sharks, we find that the surgery needed to carefully implant the strain gauges on the skeleton disrupts the surrounding muscle, leaving us unsure whether we have changed the way the shark moves. What's more, the measurements remain somewhat crude: Elizabeth Brainerd and Bryan Nowroozi of Brown University have used real-time CAT scans to show that many of the motions of a fish's intervertebral joints are subtle enough that they are still difficult to measure accurately.

WHAT'S A FISH-CRAZY SCIENTIST TO DO?

At this point the best model of a marlin backbone is not a marlin backbone. Because we couldn't study it any further in the living fish, we were left with three choices. One: quit and do another project. As depressing as that sounds, sometimes it is the only practical alternative. In the hopes of finding a species that works really well for answering a ton of different questions (which would make it a "model organism"), switching species is a common response. Two: try to build a new instrument or experimental procedure to answer the question. For the stubborn and electromechanically minded, this is often a way to work out your frustrations and keep busy while you come to grips with the fact that you really, truly are stuck. Three: build a model of your fish. For those of us who need to keep writing papers so that we can earn tenure and win research grants, this is the way to go—we model.

This may strike you as a cynical way to have backed into modeling. It is, I admit it. So we might as well go through the front door, with a smile, by asking again why a biologist would use robots to study animals. The positive answers are both practical and theoretical. On the practical side we've seen that we reach limits with both our instruments and our animals. On the theoretical side some argue for what is called a synthetic approach, a bottom-up philosophy borrowed from engineers that stands in contrast to the biologist's usual reduce-and-analyze methods: if we can build it, then we understand it.

THE SYNTHETIC APPROACH OF EMBODIED ROBOTICS

This synthetic approach underlies what Rohlf Pfeifer and Christian Scheier call embodied cognitive science or embodied artificial intelligence, and it is at the core of the defense of robots I offered earlier: build embodied robots that behave as autonomous agents. The behavior that these agents create can then be understood on the basis of their physical design, programming, and interaction with the physical world. If we can build it, then we understand it.

I should mention that although this synthetic approach with embodied robots is new to biology, physical models have been used to great effect for some time. Vogel, the biomechanics professor from Duke University, pioneered the use of physical models to test ideas about which engineering principles nature is exploiting in organisms. One of Vogel's and Wainwright's former students, Mimi Koehl, a biomechanics professor at the University of California, Berkeley, is world famous for building physical models of animals—both living and extinct—to test ideas about the functional principles in operation. Physical models have a lot going for them.

1. You can build a simplified version of an organism or its part.
2. You can enlarge or reduce the size of the part or the organism.
3. You can isolate and change single parts, keeping all else constant.
4. You can reconstruct extinct organisms.

For Koehl physical models complement experiments with both real organisms and computer models. Although we've talked about the limits of experiments on real organisms, like marlin, it's worth mentioning briefly here some of the problems with computer models. As many of us have learned, computer models are fantastic when you can represent the biological phenomena you are modeling in well-formed equations or even clunkier but still serviceable numerical recipes. Beautiful is the clean-line output, perhaps cloaked with a surface function painted in a million hues of color, of a computer model to the abstract-art–trained eyes of a scientist. But to get that elegant output, we always have to make, even in the most accurate models, many simplifying assumptions. The trick is to make the right ones.

My first step when experiments with marlin left me at the end of my conceptual tether was to create a computer model. From my biomechanical tests I had derived equations of motion that described the torque needed to bend each joint and the resulting angular motion. My initial assumption was that these equations were sufficient to describe the mechanical behavior of the backbone in a swimming marlin. I simplified the backbone mathematically to a series of equations that were linked by the bending torques that we imposed on one and then another joint. I assumed that muscles and water resistance were acting to create the torques being transmitted up and down the backbone. With these assumptions and simplifications, I created a beautiful animation of an undulating backbone passing waves of bending from head to tail.

I presented this model to the department as the capstone of my PhD research. After the talk a fellow graduate student, Matt Healy, hurried up to me and said, in a concerned hush, "You've got a problem. I just heard Vance Tucker say that you may have violated the laws of physics." This was the equivalent of saying, "A god-like scientist thinks you've made a huge mistake and your reputation and career are in immediate jeopardy"—gulp. Tucker, a physiologist working in the physically messy world of bird flight by using brilliant flow-tunnel experiments and engineering theory, was another one of Duke's biomechanics gurus. I immediately shuffled to his downstairs office.

In response to my knock Tucker looked up from his lab bench and, with a flash of recognition, said, "Come in." Terse understatement—bad sign. *I'm in worse trouble than I thought*, I said to myself. I sat down and, unable to bear any longer my impending professional death, pointed right to the sword of Damocles hanging over my head, saying, "I heard that you think I violated the laws of physics?" His response was gracious and careful. Understanding the close proximity of his criticism to my forthcoming dissertation defense, he reminded me that the model I had presented was but one of five chapters in my thesis. Its problems were, by themselves, unlikely to overturn any experimental results because the experiments were independent of the model. And, he offered, he hadn't seen for himself the model chapter. "But," he finished, "it appears to me that you've created a perpetual motion machine."

Tucker was right, and I knew it immediately. My various assumptions and simplifications allowed me to create a model that violated the Second Law of Thermodynamics. I had assumed and simplified away energy input and loss so that my backbone, once bent, would continue undulating away forever.

It turns out that this kind of violation of reality is not so uncommon: years later a master of physical modeling and bioinspired design, Charles Pell, would say to me, "Every computer model is doomed to succeed." Any computer modeler can always create beautiful output even if the physics of the model were wrong. This giant pitfall of the naive computer modeler (ahem, yes, that was me) is the reason that researchers like Pell, Koehl, and Vogel are careful to build physical models. As Pell said to me at another time: "Physical models can't violate the laws of physics." If an engineer's design violates the laws of physics, the machine won't go on forever: instead, it just won't go. So we now have a fifth reason to use physical models and not digital ones to understand biological systems. This reason is so important, however, that I will make it point number one:

1. You can't violate the laws of physics.
2. You can build a simplified version of an animal.
3. You can change the size of the animal.
4. You can isolate and change single parts, keeping all else constant.
5. You can reconstruct extinct animals.

Do not read this, please, as saying that all computer models violate the laws of physics. Many, many computer models accurately model the physics of the world. You just have to be careful and skilled about which simplifications and assumptions to make.

For my part I've realized that the mathematical representation of the biology and physics of swimming fish is, as some say, "nontrivial." In fact, for a long time many research labs around the world have been working on the hydrodynamics of flexible bodies, like fish, interacting with a surrounding fluid. The applied mathematician James Lighthill was knighted in 1971 for his efforts, which, among other things, describe how a fish creates thrust. Today, teams of biologists, fluid dynamicists, mathematicians, and computer scientists attempt to couple the physics of fluid with that of muscle and connective tissue. Nontrivial, indeed. Robert Root and Chun Wai Liew, both of Lafayette College, and I collaborate on this front, and because of their expertise, I am happy to report that I'm no longer accused of building computer models that violate the laws of physics. In terms of how they represent the physical world, the computer models that we make are not as complex as a robot. Although our computer models of swimming fish are two-dimensional, our fish-like robots are three-dimensional. Our computer models of swimming fish have a lower speed limit, below which Lighthill's thrust equations don't work; our fish-like robots can slow down and even stop. Further proof of Rodney Brooks's dictum: "The world is its own best model." If you want to be sure that your model hasn't left out any important physics, the best thing to do is to build it in the real world.

We can now revise our list of reasons to use physical models, here adding reasons based on what we know about autonomous robots. With physically embodied robots built to model animals,

1. You can't violate the laws of physics.
2. You can build a simplified version of an animal.
3. You can change the size of the animal.
4. You can isolate and change single parts, keeping all else constant.
5. You can reconstruct extinct animals.
6. You can create animal behavior from the interaction of the agent and the world.
7. You can test hypotheses about how animals function in terms of biomechanics, behavior, and evolution.

RECONSTRUCTING THE PAST

Now you can see why I got interested in physical models to study backbones. But there is one point I have completely ignored so far: the ability to reconstruct the evolution and behavior of extinct organisms.

Consider the complexity of the marlin backbone. It is unusual enough that Block mentioned it in her pitch to get me working on the organism. The point is, of the fifty-eight thousand vertebrates, if we looked at a variety of species, you'd see a great diversity of backbones. Some species have a continuous collagenous rod lacking bones, called a notochord. Some have a series of vertebrae, bones that form the vertebral column. Some have something in between, with what looks like partial vertebrae forming around or along the notochord.

What's more, we know from the fossil record that our earliest vertebrate ancestors lacked vertebral columns themselves, instead having only a notochord. This continuous axial skeleton evolved earlier in a group of animals known as the chordates. In addition to vertebrates, chordates include living nonvertebrate species, like sea squirts and lancelets. From some group of long-extinct, notochord-bearing chordates, the first vertebrates arose over 530 million years ago. There must have been some problem that being a vertebrate and then having bony vertebrae solved. The question was, what?

I first got into this evolutionary question when I was studying blue marlin. At the time, in the lab of Serge Doroshov at the University of California, Davis, I was studying white sturgeon, big freshwater fish that keep the ancestral notochord as their backbone, even as adults. I would film living ones and subject the backbones of dead ones to the same tests I was using on marlin backbones. The basic hypothesis was simple as can be: vertebral columns, by virtue of possessing rigid bones, would be stiffer in bending than would notochords. Our data suggested that this was correct.

It still didn't tell us much about the why this trait in marlins evolved or why it did not in sturgeon. As Steve Vogel likes to say: "Biomechanics is about tactics, not strategy." In other words, biomechanics can tell us about the functional consequences of different structures but not why those different functions may have conferred behavioral and evolutionary advantages to the individuals that possessed them. To make the leap to having anything relevant to say about the evolution of vertebrates, I had to assume (here we go again) that what we learned from two species of fish applied not only to other species of fish but also, in particular, to ancient swimmers like *Haikouichthys*. These little inch-long jawless fish lived some 530 million years ago and had what looks like little bits of irregularly shaped cartilage blobs on and around its notochord. Revisiting an idea first proposed two centuries ago by Sir Everard Home, Karen Nipper, an undergraduate working in my lab, and I figured that increased stiffness ought to be what you need to swim faster. A stiffer backbone would be a bigger spring, storing more energy that could be used to power the tail.

Knowing that it would be terrifically difficult to measure speed and backbone stiffness in many species (just measuring marlin and sturgeon took me several years), Nipper came up with an easy proxy for backbone stiffness: the number of vertebrae. She also had to find a stand-in for maximum swimming speed, which is notoriously difficult to measure: the swimming fish's "propulsive wavelength," roughly the curviness of its body as it swims. Fish with a large propulsive wavelength, like tuna, tend to swim much faster than fish with a small propulsive wavelength, like eels. When we correlated the propulsive wavelength with the number of vertebrae, we found a weak but statistically significant relationship. As the number of vertebrae increased, the propulsive wavelength decreased. Converting this proxy-based result back into our variables of interest, we expected that stiffer backbones would allow their possessors to swim faster than those with floppier backbones.

A complementary approach, known as the phylogenetic approach, pointed us in the same direction. A phylogeny is the branching pattern of ancestor-descendent relationships that describes the evolutionary history of any group of organisms. You can reconstruct these relationships and the timing of evolutionary change by building what is known as a phylogenetic tree—a network that clusters species according to their genealogy. Strictly speaking, a phylogenetic tree is a hypothesis about evolutionary relatedness; it can be tested by collecting new data about the shared features as well as data from newly discovered features and new species. Once you have a tree that is well-supported by a variety of data, you can use it to answer questions about the pattern of evolution. You can map out related features, like notochords and vertebral columns, onto the branches of the tree. You can learn what feature came first, how many different times the feature evolved, and what other traits your feature of interest evolved alongside. This ability to map changes in features, what phylogeneticists call character state evolution, is what makes phylogenetic analysis such a powerful tool.

Using a phylogenetic tree of vertebrates, Tom Koob, a biochemist formerly of the Shriner's Hospital for Children, and I correlated the pattern of vertebral evolution with changes in swimming behavior. When you map out just the evolution of vertebrae onto a phylogenetic tree of living vertebrates, you get a big surprise: vertebrae appear to have evolved from notochords at least three times. Vertebrae convergently evolved in elasmobranchs (sharks, skates, rays), ray-finned fishes, and tetrapods (amphibians, reptiles, bird, mammals). "Convergent evolution" is a fancy phrase for the same feature—in this case, vertebrae—having evolved independently in different species. Convergent evolution excites the heck out of biologists because it is like naturally repeating an experiment and seeing if you get the same result. Convergent evolution is thus taken as indirect evidence for similar kinds of selection pressures—in different species at different times and places—causing a similar outcome. In the case of vertebrae, they appear to be a good solution to a similar evolutionary problem. But still the question: what is the problem that vertebrae solve?

Thinking fish, fish, fish, Koob and I overlaid on this pattern of convergent vertebral evolution the pattern of changes in swimming behavior. Because we really know so little about swimming speeds and accelerations in vertebrates—which is the same problem that plagued us in the biomechanical analysis—the correlation was weak and, therefore, disappointing. First off, we had to leave out the land-based tetrapods because few adult tetrapods have retained their ancestral fish-like bodies and swimming behaviors. Second, with only elasmobranchs and ray-finned fishes to compare, we only have two large points on the map. Given those caveats, what we think we see on the tree is that vertebrae are correlated with faster swimming. Observations of single species appear to bear this out: frilled sharks with notochords are slow and plodding; mako sharks with vertebrae are some of the fastest fish in the sea; paddlefish with notochords cruise along but are not acrobatic; salmon with vertebrae leap over waterfalls. We were left with the same expectation our biomechanical analysis generated: stiffer backbones would allow their possessors to swim faster than those with floppier backbones.

But this expectation—this prediction—even though it is based on biomechanical and phylogenetic data, isn't satisfying because it leaves so many questions unanswered. Are the proxies for stiffness and speed reasonable? Is the phylogenetic tree accurate? What other parts of the body, like muscles and shape, influence stiffness and speed? Do we find only a weak correlation because other parts of the species are different too? Would the correlation hold up if we could measure top speeds seen in the wild? Might stiffness also impact other parts of swimming performance, like acceleration and turning? What are the trade-offs in performance with increased speed? And worst of all, these questions don't even speak to the evolutionary question of the dynamic process of adaptation.

When we ask why vertebral columns evolved from notochords, we are asking about adaptation. For biologists adaptation is the process by which natural selection acts over generational time to alter—to evolve—the characteristics of a population of organisms. Evolution by natural selection—as proposed by Darwin and supported since his time by thousands of experimental and observational tests—happens when the following conditions are met: (1) a feature, like the backbone, varies from individual to individual; (2) genes, at least in part, code the feature and its variations; and (3) the feature's variations impact how individual organisms behave, survive, and reproduce relative to others in that population. When these three conditions are in place, what we see as we watch a population over time is that some individuals are better at making babies than are others. Because of these individual differences in reproductive output, as individuals and generations die, the population looks different, physically and genetically, from what it once looked like. This change over time is what Darwin called "descent with modification" and what we now call "evolution by natural selection."

FIGURE 1.1. **Evolving robots.** Three autonomous, fish-like robots compete with each other for food. Because the swimming mode, sensory system, and brain of these robots are based on the tadpole-shaped larvae of sea squirt chordates, we call them "Tadros," short for "tadpole robots." Each Tadro has for its axial skeleton a notochord of differing stiffness. The stiffness of the notochord controls the swimming performance of the Tadro. Stiffness of the notochord is genetically coded and can, therefore, evolve from one generation to the next.

Like a clumsy criminal, adaptation leaves behind many clues in the DNA and anatomy of extinct and living species. But adaptation never leaves behind witnesses or a surveillance tape. Biologists inevitably have to guess at the process of evolution. The best guesses about what went on come from reconstructing the events. Using the clues—the physical evidence—good investigators can piece together a step-by-step sequence of places, agents, and interactions that most likely caused the outcome.

And what can we do to test this sequence? We can build models, let them run, and see if their behavior matches our predictions based on our evolutionary reconstruction. But we can also do one better: let the models evolve. This idea is what would ultimately lead us to invent something my students, collaborators, and I came to call Tadros (Figure 1.1). Starting with those little autonomous robots—not much more than a small computer in a bowl—we were about to embark on a journey of considerable discovery that would help us understand not just what a backbone does for a marlin, but what evolution can do for technology, and what technology can do for our knowledge of the history of life. Which is to say, Tadros themselves would be the best answer to the question: what do robots have to do with biology?

chapter 2

THE GAME OF LIFE

"**G**REAT IS THE POWER OF STEADY MISINTERPRETATION." This lament by Charles Darwin, from his sixth and final edition of *The Origin of Species* in 1872, summed up years of simmering frustration. Many of his critics and even some of his well-meaning champions had oversimplified his particular theory (other theories existed at the time) of descent with modification, what we now call evolution. The oversimplification was this: descent with modification has a single cause, natural selection.

Although natural selection was Darwin's most important insight, he recognized and stated repeatedly in print that while it was the primary mechanism of change, it was not the only one. "Evolution by natural selection" was the phrase that I used in the previous chapter to define "adaptation." Though that may be true, it's only part of the evolutionary picture. Oops. Because I didn't talk about other kinds of causal mechanisms—like mutation, recombination, genetic drift, and assortative mating—I'm one of the oversimplifiers. Let me make amends here to get you ready for the lifelike complexities of evolving robots.

I think that Darwin, a keen observer, would've loved watching our evolving robots. With them we can show what evolution looks like when selection is dominant, on the one hand, and when it takes a backseat to other evolutionary mechanisms, on the other. We can use robots to look at evolutionary processes, those ongoing, real-time, cause-and-effect interactions of autonomous agents with their environment—at any specific place and time. That is, we can become spectators at the greatest game on Earth: the game of life.

Think of it this way: life is a game, a never-ending contest played on the world's stage. But the players are not often locked in open combat. Although a great white shark hunting a California sea lion makes for dramatic theater on Animal Planet's "Shark Week," in the real game of life most of the players never meet. Instead, each player is more like a plodding decathlete, doing ten different sports in quick succession and often at the same time. Each mobile, autonomous animal navigates its landscape, finds food or hunts for it, figures out how or if to eat what it's found, detects and escapes threats, seeks and selects mates, finds shelter if it can, and makes offspring. Winners are those who survive long enough to reproduce. Among the winners, the champions are those who have the most children. The game of life is called evolution, and robots are allowed to play.

RULES OF PLAY

Evolution is one of the simplest games on the planet. It has only three rules:

1. You score points for each child you create.
2. You score bonus points if your children make offspring of their own.
3. You can use any means to make children and to help your children make children.

Just because the rules are simple doesn't mean that the strategies are simple too. Some players may realize that cooperation provides

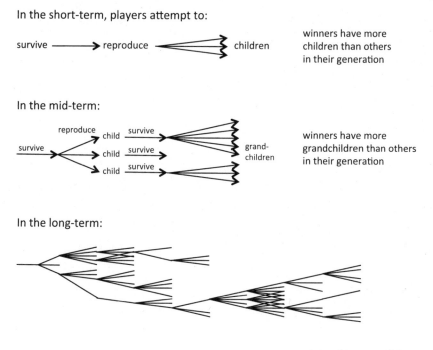

FIGURE 2.1. *The game of life: evolution.* The goal is to stay alive long enough to reproduce, to reproduce more than others in your generation, and to make offspring that are, in their own generation, successful reproducers. Winners and champions are recognized and distinguished by how well they do relative to others, not on an absolute point scale. As generations pass, results can change if short-term winners leave few descendents in the long term.

more eyes, ears, and noses for collecting food and avoiding predators.[1] Other players may figure out that if they don't have to raise their offspring, they can spend much more time making them. Some may discover that, through deceit or cuckoldry, they can have others raise their children. Others, still, might focus on collecting and protecting the resources that they need to raise their children. Players may also figure out that selecting the right mate can make the difference between success and failure.

In the real game of life most players, even humans, don't evaluate the evolutionary effectiveness of their behavior. How can we, when most of us don't even know that we're playing the game? Instead, the game is played through instinct, gut-level emotional reactions to the circumstances that present themselves. Even if you know that you're a player, know the rules, and know what it takes to win, you can only guess as to what might make your children successful at making grandchildren. You can't see the future and the chance events that may alter the conditions on the playing field. We are playing blind.

Although you as an individual may not score points through the three rules described above, you are, in fact, part of a team, and you get some points for just being part of a team that has players that are reproducing.[2] You can help other members of your familial team reproduce or raise children. But please don't misunderstand: if you are personalizing this, thinking about your own play in the game of life, then you may be getting bummed out right about now, because I may have conveyed that without children and grandchildren you are a loser. That's not my intent. I'm just trying to show you a different way to think about evolution—by thinking of it as "differential reproduction."

Understanding the fundamentals of evolution is key because we so often get it wrong. Think of Darwin's lament. Evolution is part of our everyday parlance, and even though the game of life is a fact of life, we intentionally and unintentionally misrepresent, steadily. We think, incorrectly, that individuals evolve, that individuals act for the good of the species, that some species are primitive and others are advanced, that a ladder of life describes descent with modification, and that evolution is always working to make species better. We incorrectly intuit that complexity is always more evolutionarily advanced than simplicity, that evolution is goal driven, that evolutionary change is linear and in one direction, that any anatomical structure evolved long ago for the function it fulfills today, and that humans aren't evolving anymore—wrong, wrong, wrong!

These same poison apples will tempt us when we talk about evolving robots. Our intuitive, mistaken expectations will bear fruit in the form of disappointment, disapproval, and denial. So keep this in mind: nowhere in the rules of the game of life does it say that the players in the future will always be smarter or better in any way than those playing the game right now. The rules don't say that different strategies that work at one time and place will work at other times and places. And the rules are silent regarding the behavior of other players toward one another, the number of players, the kinds of players, the availability of resources, the use of the environment, and accidents and other chance events that may occur. When robots evolve, we simply don't know what will happen. That's life.[3]

INDIVIDUALS ARE SELECTED BUT DON'T EVOLVE

I have more bad news: individuals don't evolve. As much as it makes good science fiction when Captain Jean-Luc Picard "devolves" into a lemur on *Star Trek: The Next Generation*, individuals are trapped in time and space. An individual human carries a genome—the total complement of genes and DNA within that individual—that is the product of genes from mom's egg and dad's sperm. This new combination of genes interacts with the world to create an embryo, just like our autonomous agents interact with the world to create behavior. The behavior of the genome, in this case, creates the ongoing interaction with the world known as development, the splitting of one cell into two, two cells into four, and so on, creating a multicellular animal from a single cell in a matter of a few hours.

The inside of each cell—with its aquatic world of chemicals in solution, lattice-like network of tiny skeletal structures, and membrane-bound micromachines—is the world of the genome. The genes interact with proteins that wind up and organize the DNA; the genes interact with other kinds of proteins that signal when the gene should make RNA; the DNA interacts with itself in order to make copies prior to cell division.

Each cell also has a world with which to interact. Other cells cling to it, pull on it, and exchange chemicals and electric charge. Fluid not in cells can be present in some tissues, and that extracellular fluid can bring hormones that the cell reacts to, setting up a cascade of molecular signals that results in changing how the genome is working. In response to being in different positions in the multicellular embryo, some cells "differentiate"—that is, their genes start making different kinds of RNA, and the RNA starts making different kinds of proteins. Those different proteins self-assemble into different structures. Quickly, you have cells in a particular neighborhood of the embryo that are working together to make a notochord, the skeletal rod that runs from head to tail in all vertebrate embryos and is retained in the adults of some fish and amphibian species. The cells making a notochord, in turn, release compounds that cause neighboring cells to start making a central nervous system and so on, throughout the lifetime of that individual: the genome, copied and partitioned into cells, interacts with its local world inside the cell, outside the cell, and embedded in differentiating tissues that are, in turn, interacting with the world outside the individual.

How, given all this developmental interaction of the genome and the world, is an individual trapped in time and space? Each individual is literally a product of their time and place (where time and place = "world" as I'm using the word). Take the same genome and put it in a different time and place, and you will get a different individual. Interaction of the genome and the world unfolds in development, and development reflects that particular history of that agent. Each agent is "trapped" in the sense that their agent-world interactions—which are unique—make them what they are as they continuously become what they are.

Sounds a lot like a self-help manual, eh? That self-help, you-can-change, Zen-transformation approach to our own individual histories leads us to equate our own development with evolution. That's the crux of the problem for our intuition. While both development and evolution are change-over-time phenomena, what changes in each process is different. Allow me to oversimplify: in development it's not the genome that's changing but rather what the genome makes, the material substances of the individual. In evolution it's the genome that changes, and in spite of the fact that we can have cell-level mutations that make different copies of the genome within an individual, the only way for the changes in the genome to have an evolutionary impact is for those changes to occur in egg or sperm and be passed on to the next generation.

NATURAL SELECTION EVOLVES POPULATIONS

Here's what natural selection looks like. Individuals in a population coexist in time and place. Individuals differ in their anatomy, physiology, and behavior. Openly or unknowingly, individuals cooperate and compete with each other for sex, sustenance, safety, and shelter. Some individuals are better than others at cooperating and competing. Differences in anatomy, physiology, and behavior cause some of the differences in cooperation and competition. Thus, differences in anatomy, physiology, and behavior endow some individuals with an advantage over others in the game of life and the struggle for existence. Those advantageous differences enhance some individuals' ability to survive and make offspring. That's natural selection.

If those advantageous differences can be passed on to offspring—meaning that those differences are encoded in part by genes—then the next generation will look, function, and behave differently from the previous generation. That's evolution by natural selection.[4]

As you can see, with natural selection, as we've just defined it, some individuals wind up having more offspring than other individuals do (look back at Figure 2.1). Individuals that out-reproduce others are said to have been "selected for," and those that lose in the game of life are said to have been "selected against." In this sense, every single individual of the same species, together at the same time and place—a grouping that biologists call a "population"—is "under selection" if the conditions described above hold.

So the bottom line is this: individuals can be selected, if the conditions are right, but they don't evolve. This selection means that only some of the parents reproduce, so each successive generation looks *as a group* very different from the parental population *as a group*. The group, the population moving through generational time, is the entity that evolves. Sorry, you rugged individuals, but that's the way the game of life is played.

MAKING A DIFFERENCE

Because you, as an individual, don't evolve, passing on your genome—making babies—is the best you can hope to do in the evolutionary game of life. Individual life-forms can make babies in two ways. They can make multiple copies of themselves that have nearly identical genomes, a process that biologists call cloning, or asexual reproduction. Individual life-forms may take a second path, sexual reproduction, in which the individual produces eggs or sperm, known collectively as gametes, and engages in some process to put their gamete in close proximity to another gamete from the same kind of life-form. Most plants and animals reproduce sexually. Plants do this, as your parents told you, with flowers and pollen, sometimes with an animal, like a bee, acting as the intermediary. Animals reproduce sexually either by spawning or by depositing gametes in their partners.

Generally speaking, sexual reproduction brings together the genomes from two different individuals into one new individual; it is thought to be better than asexual reproduction in producing offspring that are variable (although some plants can fertilize themselves). Both asexual and sexual reproducers can also have mutations—changes in the genetic code—that can be passed on. To be passed on, the mutations have to occur in the cells that will make the offspring. For sexual reproducers that means mutations have to occur in the cells that create gametes. The making of gametes, a process known as gametogenesis, has several important features. One is that each gamete gets only half the parent's genetic material, one from every pair of chromosomes (in humans, most cells have twenty-three pairs of homologous chromosomes, for a total of forty-six, and gametes have just twenty-three unpaired chromosomes). Another is that during gametogenesis a process known as *crossing over,* or recombination, occurs; essentially this means that the genetic material from one chromosome in a pair is shuffled to the other before the pair is split up and delivered to separate gametes. The result is both new (mutated and/or recombined) genes, and new combinations of genes in every gamete produced.

Sexual reproduction's secret weapon is the final twist: bringing together sperm and egg. When sperm and egg meet, they create a single cell, called a zygote, which has half the genome of each parent. You can see right away why offspring from sexually reproducing parents are different and why sexual reproduction is such an excellent means of producing the variable populations required for evolution by natural selection to happen.

MEASURING EVOLUTIONARY CHANGE

You've got enough information now to figure out how you can detect evolution in action. Think about measurement. What could you measure? Keep in mind that you've got to measure features of the population. You need to sample individuals and claim, usually with statistical reasoning, that the individuals you sampled represent the whole population. Or better yet, measure every individual in the population, as Rosemary and Peter Grant have done with ground finches on the island of Daphne Major in the Galapagos.

If you head out to Daphne Major with the Grants, you'll see that they net finches, weigh them, and quickly measure the size and shape of their bodies with a pair of calipers, which is basically a high-resolution ruler.[5] They tag each individual with colored bands so that they can keep track of them. They spend hours and days observing males and females nesting together as the birds select and process food, lay eggs, and feed chicks. They measure and tag the chicks. The mountains of data, collected over years, are then analyzed for things like the average length of the bill in that generation of birds. The Grants can then look at how the average length of bills (and many other features) changes from generation to generation. They can also measure how the variability, what statisticians call variance, of the length of the bill changes over generational time.

When the average and/or variance of the length and thickness of the bill changes from one generation to the next, it is the first clue that natural selection and other evolutionary forces are at work in this particular population at this particular time and place. These visible physical and behavioral features of the birds are what biologists call "phenotypes." Any phenotype may or may not have a genetic basis. If bill length has, at least in part, a genetic basis, then the change in the average bill length over generational time is evolution. The change in the average and variance of a phenotype within a population is one way to measure evolutionary change.

You can see, though, that we can run into trouble if we forget about our the conditions for evolution by natural selection. What if the phenotype doesn't have a genetic basis? What if individuals learn some new trick that isn't genetic? We can measure changes in the presence of the trick from generation to generation, so we think we are measuring evolutionary change, but upon closer inspection we find that the transmission of the behavior occurs by parents teaching their young how to do it. Orcas, for example, teach members of their pod how to specialize in hunting. Members of some pods eat otters. Members of other pods eat seals. Because we can observe the old teaching the young, we know that the tricks of the trade are learned rather than inherited genetically.

One way out of this problem is to focus, as many evolutionary biologists do, on the genotype. If the genes present in a population change, then we know that evolution has happened. We would do this by focusing on alleles. An allele is any particular version of a gene. If you have a gene that produces a protein, two different alleles of the gene may cause the protein to have a different shape or other properties. Every allele can be described as occurring in the population as a proportion, p, of all varieties of a specific gene. The change in p, where we indicate change using the Greek letter capital delta, Δ, is Δp

(read out loud as "delta-p" or "change in allele frequency"). This gives a quick shorthand for measuring evolution: $\Delta p \neq 0$. If the proportion of an allele changes in a population over generational time, then we have evolution in action. Game on!

To be fair, here, we ought to use the same sexy mathematical notation for phenotypic change. The population's average (what statisticians call "mean") value of a trait is represented mathematically by $\bar{X}$ (read out loud as "X bar" or "mean of the trait"). As long as this trait, like the length of a finch's bill, has some genes that determine it, then we have another shorthand for measuring evolution: $\Delta\bar{X} \neq 0$. If the mean of a phenotypic trait changes in a population over generational time, then we may also have evolution in action.

To prepare you for the robotic world you'll encounter in the upcoming chapters, I should mention at this point that one of the great things about creating your own evolutionary world is that you get to do things like predetermine how genes relate to phenotype. Rather than having to worry about how heritable a phenotype trait was, we just decided that genetics would control entirely every trait, X, and every variation of X. Thus, any phenotypic changes that we might see in a population would have a direct and proportional genetic underpinning: $\Delta\bar{X} = \Delta p$.

Isn't that tidy? To be careful about what we have wrought, we would say that in our population of robots any phenotypic change equals a proportional genetic change. Tidy, indeed.

There is a problem with all of this, a perceptual one: usually the $\Delta\bar{X}$ from generation to generation is so small that we, as observers, don't recognize the changes. The goldfinches in my garden this year look just like the goldfinches in my garden last year. We are blind to slow and steady changes, even those that happen over the course of a few minutes right in front of us. This phenomenon has been called "change blindness," and the fact that it happens predictably is a startling testament to the fact that we have to be told to pay attention to most things in order to notice when they change. Misdirect attention and you have a magic trick. Thus, it's no wonder that we don't automatically track the evolutionary changes happening around us all the time. For example, unless you're a gardener, you are unlikely to have noticed the Oriental bittersweet vine that has slowly crept into the shrubs and bushes of your midwestern and northeastern US yards since it was introduced in the 1860s.

Fortunately for us, Darwin—having trained with the best naturalists of the day, having traveled the world collecting samples, and having bred pigeons—was well placed to see variation and change on a small scale. Combined with his knowledge of Charles Lyell's geology, he knew that the world was old enough to have let that kind of variation build up over time to become the huge changes that differentiate whales from hippopotamuses or tuna from trout. In today's parlance microevolutionary changes cause macroevolutionary changes.[6]

Most biologists looking to measure evolution tend to focus on specific traits, or characters. John Lundberg, a biologist famous for his work on the evolutionary relationships of catfishes and the freshwater fishes of South America, told the graduate-student version of me that a character was any feature of an organism that you can observe or measure. In practice, then, you end up counting the number of spines in the dorsal fin in a population of bluegill sunfish and pumpkinseed sunfish. Or you measure whether or not the males in each species make and defend nests on the edges of the lake. Or you sample and sequence DNA to compare the alleles that make the little colored flap that sticks off the back of the gill cover. The result is that we tend to focus our analytic efforts not on the evolution of the population or species but on the evolution of one or two traits.

We do this even though we know that selection does not compartmentalize traits: the "whole animal interacting with the world and creating behavior" is really what is being selected at any given time and place, so some traits evolve not because they are the specific target of selection but because they just happen to be part of the whole animal. Changes in some traits may help the animal play the game of life whereas changes in other traits may hinder. Some changes may be neutral, but if selection on one trait is strong enough, the rest just get dragged along for the ride. For now, however, we'll just think about traits as isolated evolutionary units. To do this oversimplification, we have to perform the convenient assumption called *ceteris paribus*, Latin for "all else being equal." Under *ceteris paribus* thinking, we pretend that when we change the one thing that we are interested in, like a trait, nothing else changes or is influenced by that change. The logic of *ceteris paribus* is that we isolate one variable and understand how it influences the behavior of the whole system.[7]

We use *ceteris paribus* thinking all the time: eliminating one kind of food at a time to see if we have allergies, trying high-octane gasoline in our car to see if that improves mileage, altering our posture to see if that makes our back feel better, or testing a new drug for the treatment of multiple sclerosis in a clinical trial. *Ceteris paribus* is a great approach if all other variables remain constant and you have the discipline not to change other variables at the same time. If you remove wheat and dairy from your diet together, and that muscle soreness disappears, then you still have to go back and test each separately to know which one is causing the problem—or if it is the interaction of the two.

Using *ceteris paribus*, then, we can ask if any single trait is an adaptation. This is the equivalent of asking if a trait has evolved because it was the target of natural selection. Keep in mind that this use of "adaptation" as a noun is different from the verb of "adapting," which refers to the process of natural selection in action. In addition, if we are being careful, we'll always ask if a trait is an adaptation for a specific situation.

To answer this kind of question, we need information, and lots of it. Fortunately, Robert Brandon has carefully analyzed the kinds of information that are necessary and sufficient to provide what he calls a "how-probably" explanation of adaptation (Figure 2.2).[8] A how-probably explanation of adaptation, by the way, is rarely accomplished because we usually are missing one or more pieces of evidence. We miss loads of evidence when we are dealing with adaptation in extinct life-forms, and then the best we can do with our partial set of information is to claim that we have a "how-possibly" explanation.

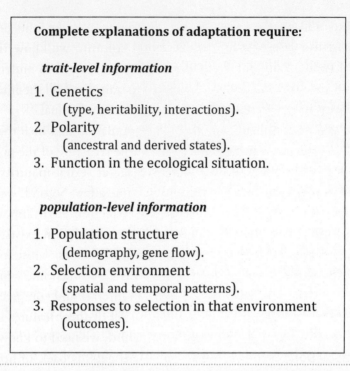

Complete explanations of adaptation require:

trait-level information

1. Genetics
     (type, heritability, interactions).
2. Polarity
     (ancestral and derived states).
3. Function in the ecological situation.

population-level information

1. Population structure
     (demography, gene flow).
2. Selection environment
     (spatial and temporal patterns).
3. Responses to selection in that environment
     (outcomes).

FIGURE 2.2. *Got adaptation?* To show that natural selection created a trait—in other words, that the trait is an adaptation—you need hard, physical evidence. You need to know about the trait and the population of organisms in which the trait exists. Collecting all of this information is difficult enough when we have the population right in front of us. Doing so when the population is extinct is impossible because we can't dig up the genetics, population structure, or selection environment. The beauty of simulating evolution with autonomous robots is that we can choose the genetics, population structure, and the selection environment. Once those features of the trait and population are chosen, we can then put our robotic population in motion and watch, over generational time, as the population evolves. Robert Brandon's 1990 book, *Adaptation and Environment*, inspired this perspective.

The evidence that we need to test a how-probably hypothesis of adaptation begins with understanding the trait of interest. First, we need to know that the trait is heritable, how it is genetically coded, and how it interacts, at the level of DNA, with other heritable traits. You can see how this evidence fits in with the definition of natural selection from earlier in this chapter. Second, we need to understand "polarity" of the trait—that is, what did the trait evolve from? What did the trait look like in its ancestral form, and what does it look like in its derived form? In the case of a jointed vertebral column, we know that it evolved from an unjointed notochord. Third, we need to understand how the ancestral and derived forms of the trait—and all the intermediate forms in-between—functioned in a living individual.

We also need information about the population in which the trait is evolving. First, we need to know about the structure of the population, things like number of individuals, age at sexual maturity, and rates of immigration and emigration, to name a few. Second, we need to know about what Brandon calls the "selection environment"— what I think of as the world in which the population exists. This world includes both physical and biological factors. Most importantly, the world contains other individuals very much like you, and because of that similarity, you are likely to interact and compete with those other members of your population. All of these features in the world make up the "selection pressure." Third, we need to know how the population responds to selection. This gets us back to how we measure evolutionary change, with $\Delta \overline{X}$ and Δp.

If you can muster all of that information, you have what Brandon considers to be an "ideally complete" explanation of adaptation. But you can see the problem with these how-probably explanations: you basically need to know everything there is to know about the trait and the population! This is what makes the Grants' work on the ground finches in the Galapagos so impressive: they have over twenty years of data on the genetics and function of multiple phenotypic traits and over twenty years of data on the demography, selection environment, and responses to selection of the population of medium ground finches on the island of Daphne Major.

Keeping Brandon's necessary and sufficient information in mind (Figure 2.2), you can see that one of the brilliant decisions that the Grants made was to select a population that was isolated (very little immigration and emigration), small, and in a simple selection environment (open habitat with only a few other animal and plant species). As Wake Forest University's David Anderson, another bird expert working in the Galapagos says, the birds on those geologically new and ecologically simple islands suffer out in the open.

What Anderson means by "suffering out in the open" is that humans who spend the time to observe carefully in the Galapagos can actually watch many events that have huge evolutionary impacts. For example, Anderson watches in lean years as Nazca boobies can only produce a few or feed some of their chicks. Reproductive success or failure is there, out in the open, for him to observe.

Make babies and help them make babies. If you are a Galapagos finch and you do this better than other Galapagos finches, then you are a winner in the game of life. Your score is based on how well you do relative to others in your population. If you are the best, you get a score of 1.0. If you are the worst and don't produce any offspring, you get a score of 0.0. This score is called your "evolutionary fitness."

Scoring the game of life is just the beginning. Once you have the score, the natural question to ask is, why do some individuals play the game better than others? And then, what about the individual and its interactions with its world matter? When you can answer these questions, then you've got a handle on which traits are important, how those traits function, what in the world selects individuals, and how the population responds to those selection pressures.

Anderson and the Grants were both lucky and smart—they managed to find an environment in which this scoring is, if not easy to do, at least possible. Most biologists don't have this advantage. Thanks to our decision to study evolving robots, my colleagues and I suddenly found ourselves in a position a lot like that of the biologists studying Galapagos finches: we could watch a population that suffered out in the open. We can create our own simplified world, create individuals whose genetics we know, create a population whose structure is predetermined, and then carefully observe behavior and evolution as the individual robots interact with their world. Because we also set up what is called the "fitness function," we are also the judges of the behavior of individuals. We become the agents of selection.

EVOLUTIONARY BIOROBOTICS

The idea of evolving robots is not new to my laboratory. Stefano Nolfi and Dario Floreano brought the concept to the general academic world with their book, *Evolutionary Robotics*, which was published in 2000. From the context of artificial intelligence, cognitive science, and engineering, they helped create a framework in which researchers could harness evolutionary processes—randomness, selection, and differential reproduction—to create without their guidance new kinds of behaviors and intelligence in mobile robots.

What we've done is to take Nolfi and Floreano's evolutionary robotics framework and apply it to biology (Figure 2.3). Whereas Nolfi and Floreano weren't originally trying to build biologically realistic robots, that's where we start. And the inspiration for that approach came from Barbara Webb, an invertebrate neuroscientist and behaviorist who figured out that she could use robots to test hypotheses

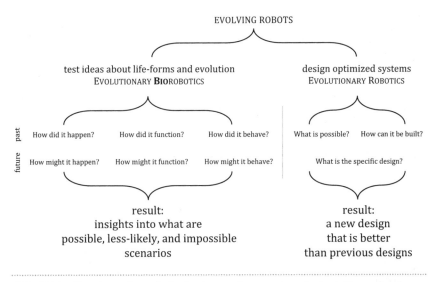

FIGURE 2.3. *People evolve robots for two main purposes: to test ideas about evolution and to design new kinds of robots.* In our laboratory at Vassar College we create evolving robots in physically embodied or digital form to test ideas about animals, evolution, and behavior. We also create evolving robots to make new designs for intelligent machines.

about the neural underpinnings of animal behavior.[9] When this approach—using physical robots to test hypotheses about biological systems—is thought of in general terms, Webb calls the field biorobotics. The combination of these two approaches creates evolutionary biorobotics.

So if we're going to build robots that can really play the game of life, they must be able to reproduce, have behaviors and other traits that are genetically heritable, and have limits placed on the number of offspring that can be reproduced. Putting these features into a robotic system gives us what we like to call the lifecycle of evolving robots (Figure 2.4).

To be frank, evolutionary biorobotics has four important limitations when it deals with extinct species and their evolution. First, as we discussed earlier when talking about the kinds of evidence that you need to explain an adaptation (Figure 2.2), analyses of past selection are fraught with potentially crippling and untestable assumptions about the genetic structure of the population; the genetics of traits in question; and the pattern, strength, and phenotypic targets of selection. Second, what you can reconstruct and test is only the ecological function of the character, the selection environment, and the response of the population to selection. Third, because we create model simulations with our robots, our reasoning is by analogy. So as we set out to explore the evolution of backbones in robotic fish, the best we could hope for was robust support—in digital and embodied populations—for the prediction that selection for swimming abilities drove the evolution of the backbone in real fish. In the worst case, the best we'd be able to say is the obvious: that different selection environments can produce different results in different robot-world systems. Fourth and finally, our use of digital and embodied robots interacting in constructed worlds grossly simplifies the animal, its environment, and the animal-environment interaction.

Still, there is much to be excited about: at the minimum, if varying our robotic backbones changes robotic behavior, at least we'd have a proof of concept that we were studying an important variable that

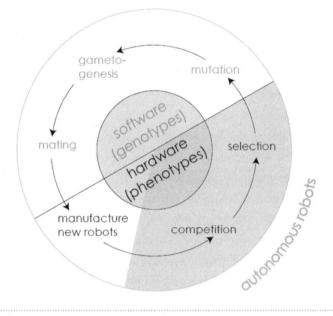

FIGURE 2.4. *The lifecycle of evolving robots.* Although all the behavioral interaction and selection in the population occurs when autonomous and embodied robots are competing (dark gray pie slice), their lifecycles also involve complex genetic interactions that occur in software (light gray font). Who gets to mate is based on evolutionary fitness as judged by a predetermined set of rules (the "fitness function"). Because the genetic interactions involve processes like mutation and mating, the genetic instructions for the next generation of robots are the outcome of random processes (mutation, mating) and nonrandom selection. One spin around the lifecycle equals one generation.

may or may not have been under selection at some point. Second, the fact that robots evolve can give us insight into how the process of adaptation works, whether in robots or biological organisms. And at least we knew we were in good company: model simulations with digital agents have already been used, most notably by Charles Ofria and Richard Lenski at the Digital Evolution Laboratory at Michigan State, to test a range of biological hypotheses about evolution.

That left us with the task of designing our first biorobots. Let's engineer some players for the game of life.

chapter 3

ENGINEERING EVOLVABOTS

"IF YOU UNDERSTAND IT, YOU CAN BUILD IT."[1] THIS IS THE engineers' secret code. It is so secret, in fact, that I can't be sure that it's real. No engineer has ever said this to me, a non-engineer, but I'm guessing that it's the last question on every licensing exam and is whispered during their secret handshake.

Regardless of whether they say it, it's definitely how they work. I figured this out for myself, having worked with many of them over the years, building robots. Engineers decide what their device should do—they understand it—and then they build it. Not surprisingly, that attitude drove me crazy—I was working and thinking in the opposite direction. At one company's early design meeting, in which the hardware, software, and mechanical engineers were pressing me for specifications, I let loose my exasperation: "Let's build the robot and then see what she can do! If I knew the specifications, then I'd know the answer. What we are doing is testing a hypothesis!" Silence. With eyebrows raised and knowing looks exchanged, the three engineers politely ended the meeting with a collective, "We've gotta get back to work."

This left me sitting with an old friend, Charles, a.k.a. "Chuck" Pell, a chief designer at the company and someone trained as a sculptor and not, I began to appreciate, as an engineer. He interpreted. Chuck explained that engineers never use the word "hypothesis," were never taught about how you test one, and were, instead, trained to build contraptions that work to do a job. The job that a contraption does, he continued, is defined by the specifications. So most engineers are literally lost without the specifications. You can't get somewhere without knowing where you are going. This all makes sense, I conceded. But it doesn't tell someone interested in evolutionary biorobotics the first thing about how to proceed. Damn the code![2]

Or don't. What I've learned since then—thanks to designers like Chuck and his band of merry engineers—is that the code provides a great starting point for any kind of design, even the crazy stuff that we do with evolving robots. In fact, implicitly, the code got us started in Chapter 2, and thanks to it, we now understand more about the game of life, evolution, and how we might go about simulating it. In this chapter we'll stick to the code and go hunting for an understanding of something more elusive—the first vertebrate. Understanding those first fish-like vertebrates—what they looked like and how they behaved over five hundred million years ago—will help us design and engineer the robotic agents that become the players in our simulation of the game of life.

NO NAME? NO GAME!

Designing evolving robots of any kind has a number of important steps. The first and the most important is not something in the engineers' code, although I think perhaps it ought to be: naming.

I feel it's my responsibility to point out that if you neglect this first design stage, if you think it's too silly to spend time on naming your robot, then you'll regret it. You'll find that other people will automatically and impulsively toss out names as they encounter and work with even just the idea of the robot. And one of those names—invariably the one that repulses you the most—will stick.

If you follow the examples of roboticists before you, you'll take one of three approaches to naming. Approach one, eponymism: name your robot after a famous person, preferably someone in robotics or artificial intelligence who is still living and can pay back the favor some day. Honda Motor Corporation took this approach when they named their bipedal spaceman-type robot "Asimo" after the great but late science fiction genius Isaac Asimov, inventor, among other things, of the Three Laws of Robotics.

Approach two, bionymism: name your robot after the animal that inspired it or the job that it does. Michael Triantafyllou at the Massachusetts Institute of Technology created the famous fish-inspired RoboTuna back in the 1980s. Bionymism, when applied to robots, often involves the creation of a portmanteau, the smushing of two words to make a new one. When smushing for your bionymistic robotic purposes, consider the common prefixes "ro-" and "cy-" along with the suffixes "-bot," "-tron," "-borg," and "-droid."

Approach three, acronymism: name your robot using an acronym that is a random letter string or, heaven forbid, an actual word related to your robot. The military loves nonword letter strings, like VCUUC, which stands for Vorticity Control Unmanned Undersea Vehicle. VCUUC, spoken as "vee-cuhk," is the serious, naval stage name of RoboTuna. VCUUC is a kind of AUV, spoken as "eh-you-vee," which stands for Autonomous Underwater Vehicle.

Now we are ready to tackle our "evolving robots." We call them Evolvabots. We really went out on a limb and smushed, using the functional variant of the bionymistic approach.

DESIGNING TO REPRESENT

Our Evolvabots need to be autonomous agents operating in an evolutionary world, but that's not all we need them to be—we need them to address our specific hypothesis, such as the relationship between swimming ability and the evolution of the backbone. In order to create those specific Evolvabots and their world, we need to ask and answer a host of mission-critical questions:

1. Which *animal* will we model and why?
2. Which *features* of the animal will the Evolvabots possess and why?
3. Which features of the animal's *world* will we model and why?
4. What is the *selection pressure* that we apply and why was it chosen?
5. How does the Evolvabot and its world, taken together, *represent* the animal and its world?
6. How will we judge if our Evolvabots are a good *model* of the targeted animal?

These questions are critical because their answers drive years of effort from a group of people, the research team. If you haven't answered these questions carefully and used them to guide your design effort, then later, when you are done running your experiments and want to get your project published in a scientific journal, you may find your team saddled with a paper that is DOA.

These mission-critical questions hark back to the "why robots?" question of Chapter 1. You have to be able to show that your Evolvabots and the processes that are used to evolve them represent, in some way, biological reality. The important word here is "represent." To represent is not the same as saying that you have to replicate exactly the actual vertebrate and its actual environment (i.e., you don't have to make a cat to model a cat). Instead, you have to demonstrate that the decisions you made in designing your Evolvabots were not arbitrary. Time, equipment, money, and expertise will always constrain those decisions. But the knowledge of your target system must also guide those decisions: you have to show that features of your Evolvabot relate to—represent—features of your target.

Representation is a general process that occurs in many different ways. For example, in biology representation occurs between the information to build the animal and the physical manifestation of the animal itself: the genome of an animal represents its phenotype (Figure 3.1). In modeling with Evolvabots, representation occurs between the robot and its biological target: the robot is a representation of the target.

How does one thing represent another thing? This is a fundamental issue in cognitive science, artificial intelligence, and philosophy of

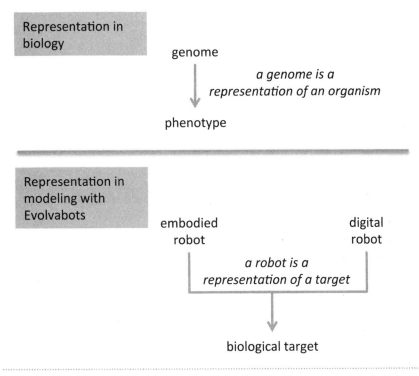

FIGURE 3.1. *Representation in biology and in modeling with Evolvabots.* In biology each animal is represented by its genome, the genetic instructions that interact over time with the environment to make the phenotype, the physical manifestation of the animal. In modeling with Evolvabots, an embodied or digital robot may represent a target, such as a vertebrate. In biology the representation is essential for development and replication of the animal. In modeling, the robotic representation is also an attempt to replicate something—in this case, particular aspects of the biological target.

the mind.[3] The most straightforward case that I can think of is when one thing is an instance of a category of things. A Tadro is an instance of an Evolvabot. As an instance, a specific Tadro represents the general category of Evolvabots. You can also flip this on its head: the category of Evolvabots represents, by definition, all instances of any kind of Evolvabot, including all the Tadros.

We encounter this kind of categorical representation all the time when we learn. Someone shows us an example of something new to us. Hey, look at this thing called a chocolate donut! Look at it. Smell it. Feel it. Taste it. This particular donut, the donut master tells you, is one example of a whole category of foodstuffs called donuts. The category, "donuts," includes other chocolate donuts that look and taste very much like this one, chocolate donuts that don't look like this one (they have sprinkles) but taste similar, and donuts that neither look like this one nor taste like it either. As you can see and taste, the representation of all donuts by a chocolate donut is created in the human mind by linking the instance at hand (or is it at mouth?) with other imagined instances. The "linking" here refers to features of the donut—looks, smell, feel, and taste—that we can morph in our minds in order to create a new imaginary instance of a donut.

So if our minds do the linking between one thing and another, and this linking is the process by which we create representations, then our mind is doing the representing. Other minds, other engines of representation, are thus the judges of our efforts at representing. If no one else thinks that we've done a good job building an Evolvabot to represent a vertebrate, then we haven't. More on judgment later.

To build scientifically useful Evolvabots, we need to use our minds and the minds of others to figure out, explicitly and objectively, how the Evolvabot represents an animal. Bloody obvious, eh? Maybe so. But keep in mind that we (meaning me and other nerds) often get so excited when we start to do cool stuff like build robots that we just start putting parts together, whatever's at hand, in order to quickly build something that works. Although this can be an exciting way to start designing robots, the implicit intuitions that guide this kind of spontaneous creation can often miss the mark in terms of clearly representing the thing that we meant to represent. So before you get started: stop! Answer the six design questions![4]

DESIGN QUESTION 1. WHICH ANIMAL WILL WE MODEL AND WHY?

We want to model the mother of all vertebrates—literally. We want the ancestor from whom all other vertebrates evolved. The only problem with this desire is that we don't know exactly who that ancestor was or what exactly she looked like. The origins of vertebrates are shrouded in mystery (soundtrack: key low Celtic whistle). What to do?

This mystery drives crazy anyone who cares about deep evolutionary history: who were the first vertebrates, anyway? This simple question turns out to be controversial because the information that we use keeps being updated and revised. Damn those meddling scientists! We find new fossils, analyze new genes, and come up with different computer methods to reconstruct evolutionary relationships among species.[5]

Some of the newest information about vertebrate evolution when we were trying to answer this question had come from the laboratory of Frédéric Delsuc at Montreal University.[6] Delsuc and his colleagues examined 146 genes in forty species of living animals, using the similarity among the genes to cluster species into related groups. The group that clustered closest to the vertebrates was the tunicates and not a group called lancelets. This result was a surprise because adult lancelets look and behave like zippy little fish whereas some adult tunicates go by the name "sea squirt" because they are little grape-like balls attached to rocks at low tide who squirt water at finger-poking people (Figure 3.2).[7] In technical terms, any two species or groups of species that are more closely related to each other than they are to any other species or group of species are called "sister taxa," where the term "taxa" is the plural form of "taxon," which means any group of related organisms.

How can it be that a bag of water is the sister taxon to vertebrates? Even though *adult* tunicates are ugly bags of mostly water,[8] the pre-adult *larvae* of tunicates look like zippy little fish, sporting a sensor-filled front end and a long tail flexing with undulatory waves that push water backward and, by Newton's third law, the larva forward. This resemblance of the larval form of tunicates to the adult form of fish has long been recognized. Walter Garstang, working in the first half of the twentieth century, proposed the then-radical idea that because the larvae of some species were more similar to the adults of others, we needed to consider the possibility that evolution might have worked by chopping off the adult stage to create new adult

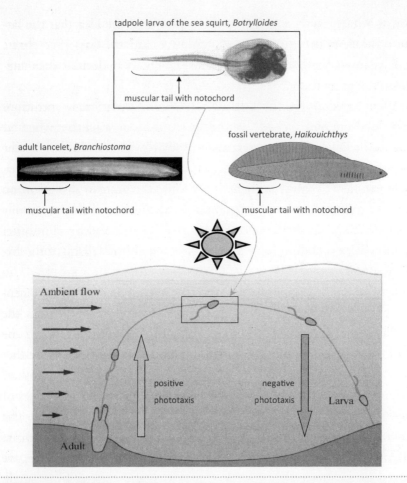

tadpole larva of the sea squirt, *Botrylloides*

muscular tail with notochord

adult lancelet, *Branchiostoma*

fossil vertebrate, *Haikouichthys*

muscular tail with notochord

muscular tail with notochord

Ambient flow

positive phototaxis

negative phototaxis

Larva

Adult

FIGURE 3.2. *Modeling the first vertebrates.* Biologists use three different kinds of animals to infer what the first vertebrates might have been like. Sea squirts (three millimeters long as free-swimming larvae of the genus *Botrylloides*) and lancelets (about four millimeters long as free-swimming larvae of the genus *Branchiostoma*; twenty-two millimeters long as adults shown here) are living invertebrate members of the Phylum Chordata, the taxon that includes vertebrates. *Haikouichthys* is a fossil fish (about thirty millimeters long) from oceans 530 million years ago and are the earliest complete vertebrates of which we know. All three animals bear a muscular tail with a notochord for a skeleton. Sea squirts have one plan in mind: swim toward the light (positive phototaxis) and away from your parent, and then swim away from the light (negative phototaxis) and find a new place to live and turn into an adult. Images of sea squirts copyright © 2010 Matt McHenry. Interpretation of *Haikouichthys* based on fossil evidence (from Wikipedia Commons: *Giant Blue Anteater grants anyone the right to use this work for any purpose, without any conditions, unless such conditions are required by law.*). Image of *Branchiostoma* licensed by Hans Hillewaert under the Creative Commons Attribution-Share Alike 2.5 Generic license.

forms. In fact, back in 1928 Garstang proposed the idea that the larvae of ancient tunicates might have provided the basic vertebrate body plan—seventy-eight years before Delsuc's molecular data suggested the same thing.[9]

Warning: sloppy-thinking watch in effect. Evolutionary intuitions may cloud inferential cognitive processes. Keep in mind that when we look at the living tadpole larvae of sea squirts, we aren't looking at the ancestor of vertebrates, even if tunicates and vertebrates are sister taxa. Living tunicates have had at least 530 million years of evolution on their own, after they split with vertebrates, to create their own family lineage. Thinking that every living species is an ancestor of another living species is a common fallacy, what I'll call the fallacy of the living ancestor.

Secondary warning: the fallacy of the living ancestor has a conceptual sibling, the fallacy of the fossil ancestor. "Paleontology is the search for ancestors," allegedly claimed George Gaylord Simpson, one of the greatest paleontologists and a cofounder of the modern synthesis of evolutionary biology. But he was wrong. (Did I just say that? Forgive me, G. G.!) The chance that you'll actually find any ancestor is very small for two reasons. First, the fossil record is incomplete. The accidental mudslides and burials that turn animals into fossils capture only a small percentage of the animals that are alive. In addition, we have good evidence that new species are usually created from small, breakaway bands of a main population; the number of these founding individuals just aren't numerous enough to be reliably fossilized and then found millions of years later. Probability of finding the actual ancestor of any living species: approaching zero.

In the end, with all of this wonderful confusion surrounding the identity of the mother of all vertebrates, the specific vertebrate we chose as our target for designing Evolvabots was the tadpole larva of living tunicates. What finally sold us was that Matt McHenry, working at the time as a PhD student in the laboratory of Professor Mimi Koehl, University of California, Berkeley, had figured out the neural circuitry involved in the swimming behavior of tunicate larvae.

Using careful experiments in which he altered the direction that light hit swimming larvae in a tank, McHenry showed that the tadpole larvae were using a very simple mechanism to orient toward and then away from the light, in a behavior known as positive and negative phototaxis, respectively (see Figure 3.2). The mechanism is called helical klinotaxis (HK) and refers to the fact that many small swimming animals move in helical pathways, as if along the threads of a screw, as they move toward or away from something in their environment, like light or the chemical plume of a food source. Although spiraling along in a helix may seem inefficient (why not just swim in a straight line?), Hugh Crenshaw, working before McHenry in the laboratory of Steven Vogel at Duke University, had shown that it was actually efficient in terms of control. To control your directions in three dimensions, all you, as a small swimmer, need to do is change two variables: your translational (straight) and rotational velocity.

When I saw McHenry present his work on tunicate tadpole larvae at a scientific meeting, I remember nearly shouting out, "Let's build a tadpole robot!" His mathematical model, which he had worked on with Jim Strother, gave us what we needed to know about the likely neural control of HK in a chordate. Fortunately for us, McHenry and Strother agreed to help transfer their knowledge of HK and tadpole larvae into a robotic form.

DESIGN QUESTION 2: WHICH FEATURES OF THE ANIMAL WILL THE EVOLVABOTS POSSESS AND WHY?

"Keep it simple, stupid." This quote, allegedly from pioneering aerospace engineer Kelly Johnson, is known throughout the design community as the KISS principle. The KISS principle is important at this stage in the design because in the heat of jubilant complexification, it helps keep your feet on the ground and your eyes on the target. KISS forces you to rephrase design question 2: what is the least we can do to fulfill our overall design goal?

For scientists, doing the simplest thing first has a very important philosophical basis: adding complexity to your model requires a combinatorial explosion of decisions, and each decision has an impact on the outcome of your design. And even more importantly, connecting back from your results to the causal elements in your design requires that you understand every element in your design and how every element interacts with all of the other elements. The simpler your design—the more KISS inspired that it is—the better your chances of understanding what the heck you've created. This KISS-first approach is one of the guiding principles at Vassar College when we work with students in the Interdisciplinary Robotics Research Laboratory. Undergraduates Adam Lammert and Joseph Schumacher, both cognitive science majors at Vassar, applied the KISS principle when they built the robots that we talk about in this chapter.

Embracing the KISS principle, we decided to keep our wish list of features short: (1) behavior: helical klinotaxis; (2) sensor: single eyespot; (3) brain: simple processor that turns the light intensity signal from the eyespot into a turning command for the motor; (4) motor: one, used for both driving and turning the tail; (5) body: a simple round bowl; (6) tail: a notochord with a flared caudal fin. Although this list may seem like a long one, keep in mind that some features are as simple as you can get (e.g., single eyespot, bowl for a body) and some features are simply missing (e.g., muscles, other sensors, a mouth).

The design of the first Tadro started in 2003 with Adam, a Vassar undergraduate and cognitive science major. He was interested in robotics, and we talked about taking McHenry and Strother's neuromuscular model for tunicate tadpole larvae and making a simple robot, relying on an insight from Chuck Pell.

Chuck had been working on three-dimensional helical klinotaxis with Hugh, of Duke University. Hugh, a biomechanist trained for his PhD by Steven Vogel, had made a true breakthrough by figuring out how to measure and mathematically describe the 3-D motion of the single-celled organisms that swim almost exclusively using HK. Later

Hugh, as a faculty member at Duke, and Chuck, working with Professor Steve Wainwright out of Duke's BioDesign Studio, created the first autonomous robot that used an HK algorithm, a small torpedo-shaped vessel. Capable of navigation with a only a single propeller for control and orientation, the robot would become known as Microhunter. For our purposes, Chuck's insight was that the three-dimensional HK used by the tunicate tadpoles would also work in two dimensions. This meant that we could stay on the surface of the water, avoiding the engineering complexities of moving in three dimensions while keeping our electronics dry. KISS in action.

For all of this work, Tadro1 was not yet an Evolvabot.[10] The transformation from biorobot to Evolvabot was driven by the interests of another cognitive science major at Vassar, Joe Schumacher, who helped endow Tadro1 with a backbone so that we could begin studying backbone biomechanics using robots.

Rob Root, Chun Wai Liew, Tom Koob, and I had tried to fund our research on the biomechanics of backbones straight-up, with no robots. We had seen two of our proposals to the National Science Foundation (NSF) rejected. The third time was a charm, and the change that made the difference—adding robots—came about almost by accident. In the fall of 2003 I worked on a review panel at NSF down in Arlington, Virginia—it was the same panel that had twice rejected our grant. The real power in the room was the program officer, who had the final say about which projects were funded. When a chorus of positivity would arise from the panelists, she'd put down her pen and start asking tough questions. When a break came in the day's work, I got a chance to ask her a question. Having previously reminded her of my two failed proposals, I went over and, without any preamble, blurted, "What about robots?" She looked up, paused without giving me eye contact, then, looking at me directly, said, "Robots would be good." That was all I needed to know.

Back at Vassar, Joe and I started scheming. Tadro1 didn't have a biomimetic notochord yet, but Adam, Tadro1's departing creator, helped Joe create Tadro2 by giving Tadro1 two important upgrades:

(1) a computerized brain (replacing Tadro1's analog circuitry) and (2) a genetic algorithm that coded for the size of a flapping tail made out of duct tape. Joined in the summer of 2004 by Nick Livingston, Joe quickly created a water world, programmed the digital brain, and set out to design a biomimetic notochord and vertebral column. For the electronics and the new Tadro body, he enlisted help from John Vanderlee and Carl Bertsche, Vassar's electronics technician and machinist.

By the time our NSF funding started in January of 2005, Joe was already replacing Tadro2's duct-tape tail with one that had a simple rod serving as a notochord. He used ten-centimeter-long cylindrical erasers as the notochord, plastic clamps as vertebrae, and then put a flared caudal fin on the end. Together we designed the genetic algorithm that would code for the evolvable traits: the length of the axial skeleton and the number of vertebrae. Nick made an important innovation: he wrote a program that allowed Tadro2 to make its tail adjustments for maneuvering using the same motor that flapped the tail. This architecture further reduced the complexity of Tadro2 and made it much more reliable. With our incoming students, whom we called "Fish Fellows," we quickly realized that notochords made of erasers weren't making sense because we couldn't change the stiffness of the erasers' material. The solution—building the notochord out of a biomaterial whose stiffness we could vary— would come from Tom Koob, as we'll see on the next pages. With the change to the brain and the tail of Tadro2, we realized that we really had a new critter: Tadro3 (Figure 3.3).

We had three reasons for thinking that Tadro3 was the Evolvabot we were looking for: (1) the brain would make it autonomous, able to behave on its own without a human "in the loop," without a remote operator acting as the eyes and brains of the operation; (2) the light-seeking behavior would emulate the phototaxis of the tunicate tadpole larva; and (3) the body would also emulate that of the tunicate larva, possessing a propulsive tail with a biomimetic notochord, a backbone whose properties we could vary by degrees, code with an artificial genome, and cause to evolve under the right ecological situation.

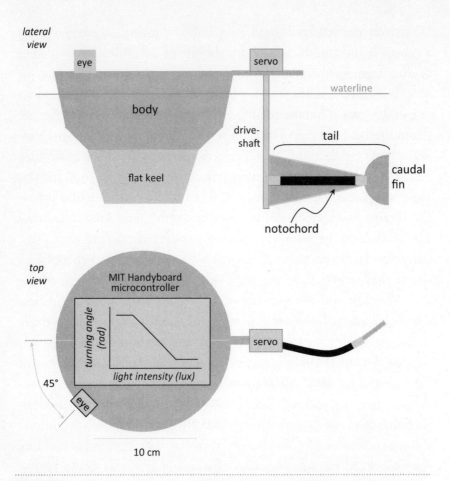

FIGURE 3.3. *Tadro3, the Evolvabot designed to represent the tunicate tadpole larva.* Tadro3 has a single eyespot (photoresistor), a flapping tail, and a microcontroller that converts the light intensity at the eyespot into a turning angle at the tail. This sensorimotor system produces autonomous phototactic navigation (see Figure 3.2). Tadro3 has a biomimetic gelatin hydrogel serving as a notochord. The notochord's structural stiffness is determined by the material stiffness of the gelatin, which we control with chemical cross-linking, and the length of the tail. Both material stiffness and length of the tail were coded genetically as evolvable characters. Proportions are drawn to scale, and more information about the specifications of the design are available.[11]

To understand more about the backbone as a feature we targeted, we need to dig a bit deeper into some of the assumptions we've been making about the evolution of notochords and vertebral columns of chordates. As I said in Chapter 1, a species called *Haikouichthys ercaicunensis*, a small, sporty little fish that lived 530 million years ago (Figure 3.2), appears to have been conducting its own evolutionary experiment on turning a notochord into a vertebral column. Widely spaced bits of cartilage or bone can be seen along its notochord.[12] The proto-vertebrae, as some authors have dubbed them, are too far apart to resemble the tightly packed vertebrae that we see in most other fossils or living species that have vertebrae. However, for all of their differences, the proto-vertebrae of *Haikouichthys* allow us to infer three important things about the evolution of vertebrae:

1. The earliest vertebrate fossils had a backbone that was primarily a notochord, supporting the contention that the notochord is the ancestral state of the vertebrate axial skeleton (no one is surprised by this, by the way, because evolutionary trees have long inferred this pattern, as we'll see in a minute).

2. Vertebrae, even though they appear early in vertebrate evolution, take millions of years to evolve into what we now recognize as a vertebral column. Philippe Janvier, a paleontologist specializing in the earliest fishes, estimates the origin of an internal skeleton of calcified cartilage or bone at about 443 million years ago, about 90 million years after *Haikouichthys*' experiments.

3. Because the backbone of *Haikouichthys* does not have the large vertebrae and thin intervertebral joints that we see in living fishes, but just the opposite, we need to be careful to recognize that the two states of the axial skeleton, notochords and vertebral columns, really demarcate the ends of a spectrum of possible axial skeletons. With that in mind, we'd expect to see throughout living and extinct vertebrates variations in the size, shape, and number of vertebrae and intervertebral joints.

Phylogenetic analysis gives us another clue about the polarity of the states, or spectrum of states, of the axial skeleton. The notochord, without any signs of vertebrae, is possessed by both tunicates and lancelets (see Figure 3.2). If, as Delsuc's tree showed, tunicates are the sister group to vertebrates and lancelets are the sister group to tunicates + vertebrates, then the simplest, most parsimonious explanation is that notochords evolved in the common ancestor of all three groups, well before the vertebrates split off and began to evolve the vertebrae that we think we see in *Haikouichthys*.

Additional evidence for the notochord evolving first is that it also appears first in the development of living fish, prior to the formation of vertebrae; vertebrae are then built in and around the notochord.[13] Although being first in development isn't, by itself, evidence for evolutionary polarity, the notochord is a central structure in early embryo development, one that is necessary for the formation of the nervous system and the growth of the embryo. Every vertebrate embryo grows a notochord first and then, if they grow one at all, a vertebral column. This invariant pattern of the notochord guiding the embryonic development of vertebrates and their vertebrae is consistent with the hypothesis that notochords evolved before vertebral columns.

In development and evolution the axial skeleton functions to stiffen the body. As we talked about in Chapter 1, stiffness is the mechanical property that dictates how much a structure changes shape—lengthens, shortens, twists, or bends—in response to having forces applied to it. Put a rubber band on your two index fingers and apply a tensile force to it by increasing the distance between your fingers. The rubber band, at least at first, lengthens easily. Now do the same thing with a shoelace, the ends of which you hold between index finger and thumb. The shoelace does not lengthen much, even if you apply as much force as you can. In engineering terms the shoelace is "stiffer in tension" than the rubber band.

Bending or flexural stiffness of the notochord can be increased by adding vertebrae.[14] Working with Tom Koob and Lena Koob-Emunds at the Mount Desert Island Biological Laboratory in Salsbury Cove,

Maine, we analyzed hagfish, a group of eel-like fish that never evolved jaws and retain, as adults, a fifty-centimeter long notochord. After a hagfish died, we removed and bent its notochord to measure the notochord's flexural stiffness. We then threaded onto the notochord, like pearls on a string, a series of rigid plastic rings that snugly fit the notochord. Sometimes we added just a few rings, widely spaced like the vertebrae of *Haikouichthys,* and sometimes we added more, leaving less space for bending. The result? More vertebrae created an axial skeleton with increased flexural stiffness. With this in mind, in Tadro3 we allowed bending stiffness itself, rather than number of vertebrae, to be the character that was genetically coded to evolve.

This may seem bass-ackwards, I admit. Why not just build an artificial notochord and then add plastic rings, as the game of life demands, to model the number of vertebrae? Our rationale for evolving bending stiffness as a proxy for vertebrae went as follows. If you evolve only whole vertebrae—they are either present or absent—then your resolution is limited to those stepwise changes. You can't see what "half" a vertebrae looks like. But do half-vertebrae evolve? Yes, sometimes. In the fossil record for the group of fleshy-finned fishes that were the outgroups to the first land-living tetrapods, we see partial ring vertebrae, little crescents of bone that cup the bottoms of the notochord.[15] At least in this group, it looks like vertebrae form from different pre-existing centers of bone formation, in this case the ribs. Sindre Grotmol and his colleagues at the University of Bergen, Norway, have shown a similar process in the development of living fishes.

Here's the rub if, like us, you are interested in evolutionary biorobotics: how do you make partial vertebrae? We've tried, trust me. My students can tell you many a tale of working on making tails with partial vertebrae and vertebrae of various sizes and shapes. But in almost every case the vertebral column would tear (fracture, strictly speaking) at the interface between the bit of vertebra and the notochord.

Our solution, at the time, was to forget about the vertebrae and make a continuous structure, a biomimetic notochord whose material stiffness we could alter and, in so doing, alter the notochord's flexural stiffness.[16] We also realized that if you changed the length of a structure, you alter its structural stiffness: for a given flexural stiffness, a longer structure deflects more than a shorter one. Our biomimetic notochord was, in the lingo of material scientists, a hydrogel made, as I said, of collagen, thanks to Tom Koob.

When you take the powdered gelatin and add it to heated water, it dissolves nicely if you stir the pot. As you cool the mixture, the gelatin, now evenly spread throughout the forming solid, makes some chemical bonds between the scattered molecules. Pour the cooling liquid into a mold of some kind and pop it into the refrigerator. In the cold the motion of the collagen fragments slows, allowing even more bonds to form. Presto! You've created a solid from a liquid: a molded hydrogel!

For biomimetic hydrogels, we poured the hot gelatin and water mixture into an array of molds that made cylindrical rods about 10 centimeters in length and about 0.5 centimeters in diameter. Once the gelatin had set in the fridge, we pulled the rods out and then did something you wouldn't do with your dessert: chemically embalm them. Embalming, or what a biochemist would call fixation or, in this case, cross-linking, keeps tissues from degrading and, gulp, spoiling.

For our hydrogels, the mortuarial embalming agent we use is called glutaraldehyde, and it does two things. First, glutaraldehyde allows us to let the biomimetic notochords warm up to room temperature without melting—it keeps the hydrogels solid. Second, glutaraldehyde allows us to control the stiffness of the hydrogel: the more time that the hydrogel spends in the glutaraldehyde solution, the stiffer it becomes as more chemical crosslinks form between collagen molecules. Here, finally, was our method for getting any intermediate flexural stiffness that a genetic call for a partial vertebra might require.

DESIGN QUESTION 3: WHICH FEATURES OF THE ANIMAL'S WORLD WILL WE MODEL AND WHY?

The world or arena that you design for your robots is as important as the robots themselves. Thus, we have in hand one of the design principles for embodied robots expounded by Rohlf Pfeifer and Cristian Scheier: build a robot for a specific ecological niche.[17] In other words, you have to build the agent with a particular world in mind. This is obvious when we think about the difference between a fish-like robot and a dog-like robot: water versus land. But what about a fish-like robot swimming in the nooks and crannies of a coral reef and one swimming in the open ocean? If we use fish as our guides, these robots ought to be very different kinds of agents, the first skilled at precise maneuvering and station-holding and the second skilled at cruising and perhaps navigation.

The world also has other players. For evolutionary biologists, the other players are called "biotic factors" and everything else is "abiotic factors." For an individual robot, biotic factors are all the other robots and animals with which it might interact. Abiotic factors include the physical and chemical situation in which it's placed. Together, biotic and abiotic factors make up the ecological niche, here what I'm calling the stage, the modeled world, or the selection environment.

We wanted a world that, like Tadro itself, was a simplification of our best-guess of ancient reality. The ancient world for the first vertebrates was, as far as we can tell, oceanic, near the shore, and full of biotic factors like giant arthropods, trilobites, anemones, and worms with legs.[18] Obviously they all had to eat, and some of them likely competed with the first vertebrates, jostling for position at the donut store, figuratively speaking. KISS demanded we leave most of that cast of characters out.

The simple world we built was a water world, a walled tank 2.5 meters across with a single sun, limited time, and three Tadro3s (Figure 3.4). The sun was a hundred-watt flood light suspended above the surface of the water. Time was limited to three minutes for each trial. Each Tadro3 competed in six different trials, with three robots in each trial. To account for the fact that each Tadro3, even though built to be identical in every way but for their variable tails, may vary in performance, we swapped the biomimetic tails among the three Tadro3s and made sure that all possible combinations of tails and robots were

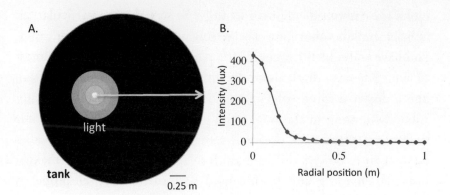

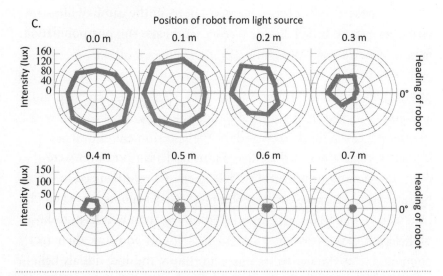

FIGURE 3.4. *The water world of Tadro3.* A. Overhead view of the surface of the water in the 2.5 meter–diameter circular tank, with the position and distribution of the light and its gradient shown by the nested circles. The big white arrow indicates a radial slice of the light gradient shown in B. C. Perception of the light gradient by Tadro3s. Polar plots indicate light intensity (along radii, with origin at 0 lux) registered by robots at different headings every 0.1 meters along the radial slice shown in B. A heading of 0 degrees means that the robot was facing in the direction indicated by the arrow in A. Recall that eyespot is located 45 degrees to the left of the robot's center line (see Figure 3.3).

tested. This swapping allowed us to make sure that no particular tail lost the game because it was always stuck with a sluggish robot.

In the water world, sun represents food. The first food, for almost all life, is the glucose made by plants that harvest the energy from light. In the ocean most critters follow the light to find the food. That's why most of the sea creatures live in the upper reaches, the shallower depths, because that's where the light is. And where there is light, there are algae and diatoms, the "primary producers" making the stuff, their fronds and bodies, that feeds the mobile, self-propelled critters like fish.

The world sets the stage for what matters in the game of life—surviving and reproducing better than other agents in your population: outwit, outplay, out-reproduce.[19]

DESIGN QUESTION 4: WHAT IS THE SELECTION PRESSURE THAT WE APPLY AND WHY WAS IT CHOSEN?

Honestly, no one has any idea what the selection pressures were that drove the evolution of the early vertebrates. As we discussed in Chapter 2, it's difficult enough to understand what is going on when you have live animals right in front of you. Even then you have to know the evolutionary fitness of different individuals and link their many phenotypic differences to differences in how the individuals behave and interact with their world over their entire lifetimes. This is a daunting task under the best of conditions with live animals. For extinct animals, all we can do is make what seem to be reasonable guesses, or what Brandon called "how-possibly" explanations.

So how do we make reasonable guesses—hypotheses—about selection pressures that drove early vertebrate evolution? We use our understanding of how specific traits function in living animals, assume the same thing was happening long ago, and then conjecture that variations in that trait had functional consequences for the individuals possessing those variations that could help or hinder them in the game of life. We then imagine the conditions of the world, with its physical characteristics and other autonomous organic agents, under which that help or hindrance would matter the most for survival and reproduction. That particular condition of the world is our "selection environment," to use Brandon's term from Chapter 2, and the "selection pressure" that we are talking about here is the particular type of interaction between the selection environment and the individuals that most affect survival and reproduction. For example, many people think that avoiding predators is the selection pressure that drives the evolution of fish coloration, body shape, and swimming performance.[20]

We focus our educated guessing on the axial skeleton and the evolutionary change from notochords to vertebrae. As explained above, Tadro3s are built to evolve the structural stiffness of their tails as a proxy for the presence of vertebrae. With tail stiffness in mind, we first think about the mechanical function of the tail: what does it do and how does it do it?

The primary mechanical function of the chordate tail, the section of the body behind the gut and including the terminal caudal fin, appears to be propulsion. No surprise. Tunicate tadpoles, sharks, and bony fish all undulate their tails to create thrust. Undulation is making waves, traveling waves of body flexion that start near the head and move toward the caudal fin. By altering the shape and speed of those undulatory waves, fish alter their swimming speed, turn, and brake.

The structural stiffness of the tail controls, in part, the shape and speed of the undulatory waves. If you've ever tuned the string of a guitar or violin, you know that if you tighten the string, it will vibrate faster when plucked, creating a higher pitch. By tightening the string, you've stiffened it, and it is a well-known principle of engineering that an elastic structure like a string or steel bridge will tend to vibrate at a particular frequency, called the natural frequency, that is determined by the structure's stiffness, mass, and ability to dissipate energy. Thus, stiffer tails ought to vibrate faster than flexible tails.

So how might have the stiffness of tails evolved? What might have been the selection pressure? Here we connect the dots. If increased tail stiffness makes or allows the undulatory waves to travel faster, then increased tail stiffness increases the speed at which fish can swim. If increased swimming speed helps fish find food, then increased tail stiffness increases the amount of food that a fish can eat. And finally, if finding and eating more food increases a fish's chances of survival and its reproductive success, then increased tail stiffness was selected to improve the ability to forage and feed.

We put the feeding selection pressure into action by coming up with a "fitness function," which is a fancy name for a numerical formula for judging how well each individual does relative to the other individuals in the population. Because foraging and feeding involve detecting the presence of the food, traveling to the food source, and then staying and eating, we reasoned that a number of behaviors should be rewarded at the same time. First, the ability to detect the food could be measured as the *time* it took a Tadro3 to reach the source; more points are given for shorter times. Second, the ability to get to the food quickly could be measured by the average *speed* at which the Tadro3 traveled; more points are given for higher speeds. Third, staying and eating could be measured by the average *distance* of each Tadro3 from the food source; more points are given for a shorter distance. Fourth and finally, the sloppiness of swimmers that waste energy, and thus food, could be measured by their average amount of body *wobble*; more points are given for smaller amounts of wobble.

DESIGN QUESTION 5: HOW DOES THE EVOLVABOT AND ITS WORLD, TAKEN TOGETHER, REPRESENT THE ANIMAL AND ITS WORLD?

Our earliest Chordate ancestors were probably little fish-like swimmers, with notochords for an axial skeleton in their body and tail and with at least a single eyespot. They could probably detect and navigate relative to light gradients in the sea. This simplified and hypothetical ancestor is inferred from what we know about living chordates and Cambrian vertebrates (see Figure 3.2), development of living vertebrates, and the evolutionary relationships among chordates that we reconstruct using phenotypes and genomic data.

Using this information, we chose as our specific biological target the tadpole larva of living tunicates. Even though no living species is the ancestor to another living species, we still felt confident enough in the similarities between the behavior of tunicate tadpole larvae and the ancient, extinct chordates to use the larva as our model for designing Tadro3. The Tadro3 uses the same neural algorithm that we think the tadpoles use, both have a single eyespot, and both continuously undulate a tail with a notochord. To flap the tail, the Tadro3 uses a single motor instead of a series of muscle cells distributed along the tail. Both Tadro3s and tadpoles turn by adjusting the angle at which the tail meets the body.

We did simplify things, however, by having Tadro3 swim just on the surface of the water rather than underneath and by building the Tadro3 on a scale easy for us humans to manipulate. Tadro3 is about twenty-five centimeters from head to tail; a tadpole larva, however, is just a few millimeters long. We also simplified, as I described above, the physical environment: Tadro3 lives in a small circular pool rather than the ocean. Feeding behavior is also simplified, as Tadro3 merely needs to navigate up the light gradient cast by the single light over the pool. Likewise, although tadpoles encounter many other animals in a twelve- to twenty-four-hour dispersal period, during which most tadpoles die, our robots encountered no other agents (just other Tadro3s), "dispersed" for just three minutes, and could not "die."

We think these simplifications were justified, but we must always be on guard against simplifications that are not. If we ever cannot justify them to other people, then we have failed in our primary goal: to test an evolutionary hypothesis of animals using robots as model simulations.

DESIGN QUESTION 6: HOW WILL WE JUDGE IF OUR EVOLVA-BOTS ARE A GOOD MODEL OF THE TARGETED ANIMAL?

Last comes the justification. Barbara Webb delineates seven dimensions that can be used to help describe and characterize biorobotic models: (1) biological *relevance*, (2) *match* between the behavior of the biological target and the robot model, (3) *accuracy* of the model in using the same functional mechanisms as the target, (4) how *concrete* the model is in terms of mimicking features of the target, (5) the *level* in the target's structural hierarchy at which the model focuses, (6) the *specificity* of the model in terms of the number of elements targeted, and (7) the *substrate* from which the model is built, either digital or physical.[21]

For Webb, the key dimensions for biorobotic models are biological relevance and substrate. You've got to have a robotic system that allows you to test a hypothesis about your target or it isn't relevant; the test of the hypothesis can be if the robotic system works as expected or, in the case of Evolutionary Biorobotics, if the evolutionary trajectory of the system is as expected. Moreover, she argues that the substrate ought to be physical rather than digital, for all the reasons outlined in Chapter 1.

For Evolutionary Biorobotics models, we include behavioral match and functional accuracy. For example, we want an individual Tadro3 to behave like a tunicate tadpole in terms of (1) using a tail that generates thrust by undulating, (2) navigating up a light gradient, and (3) being part of an interacting and evolving population. That's the behavioral match, with a three-layer nested hierarchy of the behavior of the organ, the individual, and the population.

We also want Tadro3 to use the same functional mechanisms that tunicate tadpoles do, such as the same kind of undulatory wiggle of the tail and the same neural wiring and sensorimotor loops that we've understood and engineered above. We also want the evolutionary mechanisms that we talked about in Chapter 2—selection, mutation, random mating, and genetic drift—to be what drives the evolution of the population of Tadro3s. That's the accuracy of the functional mechanisms, with a three-layer hierarchy of propulsive mechanism, sensory-neural-motor mechanism, and evolutionary mechanisms.

In sum, relevance, substrate, match, and accuracy are our primary goals for designing, engineering, and running Tadro3s as a model simulation of the earliest fish-like vertebrates. We'll judge how close we come to meeting those goals when we look at the Tadro3s playing the game of life in their water world in Chapter 4.

KISS THE CODE

Throughout this design process we've employed the KISS principle and simplified our Evolvabot and its world whenever we could. We've even argued in the scientific literature that the original Tadro1, which was built to swim and behave like a tunicate tadpole, is the simplest possible autonomous navigator because it possesses a single sensor and a single motor control output (tail offset to turn). Tadro3 has that same basic hardware and neural architecture but has the biomimetic tail coded to evolve. Simple, simple.

But as we've seen, even with a simple robot like Tadro3, you need to understand a ton of stuff about the animal target. Think: the engineers' code. Think: specifications. We had to figure out how evolution works (Chapter 2), make an educated guess about the hypothetical chordate ancestor of vertebrates, find a reasonable living proxy for that ancestor (tunicate tadpole larva), understand the swimming behavior of the tadpole, infer the neural control system of the tadpole, measure the mechanical function of a tail with a notochord, and divine a likely selection pressure on early vertebrates. Phew (pant, pant, out of breath . . .)! This is the kind of understanding that allows us to follow the secret engineers' code and build a population of evolving robots.

But were we successful?

chapter **4**

TADROS PLAY
THE GAME OF LIFE

T ADROS EVOLVE! BY ANY MEASURE, OUR POPULATION of autonomous, aquatic Evolvabots successfully played the game of life (Figure 4.1). No surprise, I hope—indeed, the surprise would have been if they didn't evolve, as that was what we designed them to do.

HOW TO EVOLVE A ROBOT: IT TAKES A SMALL- TO MEDIUM-SIZE VILLAGE POPULATED WITH EAGER SCIENTISTS

If you begin at the beginning and then go on until you get to the end,[1] you would see that from the design of Tadro1 in 2003 to the publication of our Tadro3 paper in 2006, our Tadro team had more than twenty-three members. Okay, so that's not a small- to medium-sized village. But it's still some measure of how hard this work is.

Kira Irving, Keon Combie, Virginia Engel, and Joe Schumacher led a group of Tadro3 operators and biomimetic tail benders that included Nicole Doorly, Yusuke Kumai, Gianna McArthur, and Kurt

Bantilan. They performed 120 trials over ten generations, building 360 tails with biomimetic backbones along the way. Each trial also had three minutes of videotape to analyze in one-second increments; in each frame, we had to mark (and then double-check) the position of the light and the bow and the stern of each Tadro3. We used these points to compute average speed, average wobble, time to the food, and average distance to the light source. Twelve of these numbers for each individual tail phenotype (each phenotype was described by two traits: length of tail, L, and the biomimetic notochord's material stiffness, E—more on these later) were then used to calculate the three individual fitnesses corresponding to each phenotype—fitnesses that we needed to run the genetic algorithm that spat out the next generation's phenotypes. When we were really cranking full time in the summer—including evenings—a group of four or five of us could get a generation done in about four days. But if anything went wrong, like when we dropped a Tadro upside-down in the water, then we'd have to stop and rebuild before we could get back to the evolutionary trials. All told, the trials took ten weeks.

As we let our population of Tadro3s play its game of life, we found that the feeding behavior, tail stiffness, and genetic composition of the population all changed over generational time (Figure 4.1). Although the fact that the population evolved wasn't a surprise, we were surprised about the direction evolution took. I should say directions, plural. Under a steady selection pressure that rewarded enhanced

FIGURE 4.1. (facing page) *Evolution of a population of Tadro3s.* Phenotypes and genes change over generational time. Both kinds of changes are how we measure evolutionary change. Because we know what's happening in terms of selection and random genetic events, we know when selection is a factor. The square frames around some of the generation numbers indicate when selection was strong enough to create differences in the reproduction of individuals making offspring for the next generation. For example, the change in the population's feeding behavior from generation 1 to 2 can be attributed to both selection and random genetic effects. In generations in which selection is not strong, any changes in the next generation are caused only by the random genetic effects. The population's actual evolutionary path is indicated with a solid line; its simple and expected path is indicated with a dashed line. Points are averages; error bars are standard deviations.

phenotypes

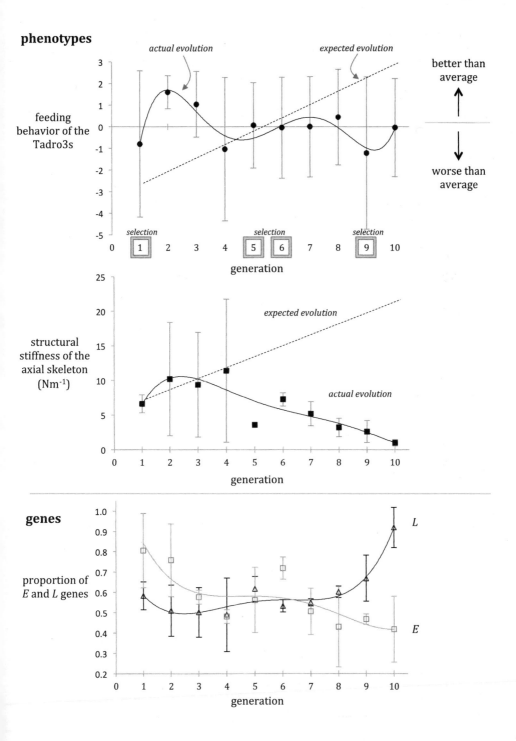

genes

feeding behavior, we expected steady, directional change. Instead, what we see, even in this simplified Tadro3 world, is that evolutionary change oscillates, moving in different directions at different times.

EXPLAINING THE UNEXPECTED

Given that we know the Tadro3 world inside and out, we sure as heck-fire better be able to explain the unexpected directions in the evolution of the population Tadro3s.

Let's jump-start the process of interpretation by revisiting the big picture. We were interested in creating Evolvabots that could test a hypothesis about the evolution of early vertebrates: natural selection for enhanced feeding behavior drove the evolution of vertebrae in early vertebrates. From this hypothesis we came up with a primary prediction: selection for enhanced feeding behavior will cause the population of Tadro3s to evolve stiffer tails. Implied in that prediction is another: the evolutionary change will be directional, moving the population from flexible to stiff tails in concert with ever-improving feeding behavior.

What's clear is that these predictions are too simple. Take a close look at the reality of the hard data (Figure 4.1). The average score for feeding behavior[2] increases greatly from generation 1 to 2 while the standard deviation,[3] represented by the length of the bars emanating from the filled circles, decreases. This initial change sure looks like the direction we predicted: enhanced feeding behavior. The decrease in variance might be expected, too, because selection was just picking individuals with the highest feeding behavior to reproduce. But in the very next generation, 3, the average score for feeding behavior drops and then continues downward through generation 4. What's up, Doc?

At first glance, this downward trend looks crazy-wrong: how can we select for improved feeding behavior and get just the opposite? The short answer is this: in generation 2 differences in feeding behavior among individuals were not large enough to cause differences in reproduction. All three individuals contributed the same number of gametes (egg and sperm)—two—to the mating pool of six gametes. Our mating algorithm assigned differences in reproductive output according to the differences in fitness between individuals. Even though the fitness function assigned slightly different numbers to each individual—based on their abilities to swim quickly, reach the light target quickly, stay and feed near the light target, and move around smoothly—in generation 2 those fitness differences were just too small to matter.[4]

When individuals contribute equally to the next generation, we have an evolutionary tie. This tie means that the parents are likely to make a generation of offspring that looks, on average, like themselves. In evolutionary terms, there was an absence of selection, or no selection pressure. Either phrase might sound a bit inaccurate, as we had selection judging individual Tadro3s in a given generation using our fitness-function scorecard. But keep in mind that ultimately it's differential reproduction among individuals that makes evolution by natural selection work.

In the absence of selection, how, then, do we get evolutionary change? Recall Darwin's lament from Chapter 2: "Great is the power of steady misinterpretation." He was referring to the fact that many scientists overplayed the power of selection, to the extent that they ignored other evolutionary mechanisms and, as a result, saw adaptations in every cranial bulge and fingerprint, when sometimes randomness is what's at work.

Darwin had little hard evidence to counter this misconception because he lacked our understanding of genetics, which, ironically, was developed by a contemporary of Darwin's, Gregor Mendel, but languished in an obscure journal until the early twentieth century. With our current understanding of genetics, of course, we know that random genetic changes are always occurring, both in the germ-line and other cells. More randomness can come into play during mating. Although some mating is decidedly not random, for many organisms it is. We set up our Tadro3 population to engage in random mating. After each gamete was given a certain probability of mutating or not, we took all six of the gametes the parents produced, and combined the gametes into pairs randomly. Those pairs of haploid gametes combined to give us the new diploid genome for each new baby Tadro3.

At this point, my colleague Rob Root, a mathematician and central collaborator on many of our Tadro projects, would want me to remind you that we have a problem with our randomness because we are operating in the realm of "the mathematics of small numbers." Our population size of three is simply too small for the statistical assumptions of flip-a-coin randomness to hold. The result, if you were flipping coins, is that you could easily hit heads, heads, and heads, three in a row. You'd say that you were "on a roll," and it sure would *not* look like a random process until you flipped the coin about twenty more times. Geneticists call this mathematical description of what happens in populations of small size "genetic drift."[5]

Genetic drift, in the absence of selection, produces evolutionary change in directions that are random with respect to both phenotype and genotype. Because drift is random, neither it nor any other random mechanism will produce the kind of long-term directional pattern that we recognize as the signal of adaptation. Only selection is equal to the task.

For our tiny population of Tadro3s, the absence of selection in some generations allows the combined random chance of two sources of randomness—mutation and genetic drift—to be the prime movers. Over a generation or two, we can even be fooled. Our population can be "on a roll" that may look like a directional and evolutionary trend. This is what happened to the average feeding behavior from generations 2 through 4. Random changes just happen to have combined to decrease the feeding behavior. Chance favors no one.

EVOLUTIONARY MECHANISMS: A TRIUMVIRATE

What's really important to keep in mind, and to keep Darwin from being on a roll in his grave, is that over generational time random changes—which occur with or without selection—are, indeed, mech-

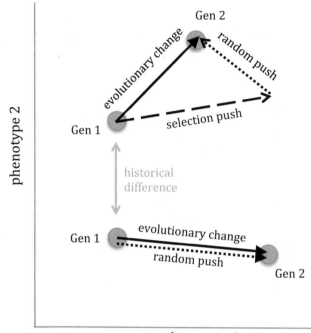

FIGURE 4.2. *Selection, random processes, and history interact to create evolutionary change in a population.* The dots represent the average position of the population in "morphospace," the realm of evolutionary possibility represented by the
ranges of phenotype 1 and phenotype 2. History determines where in the morphospace a population starts as well as the nature of the genetics underwriting the phenotypes. Evolutionary change is the movement of the population, over generational
time, in response to either (1) selection and random processes (top population) or
(2) random processes alone (bottom population). Selection is unlikely to act without
random genetic processes because the continual creation of genetic variation is
needed to make the phenotypic variations from which the fitness function selects.

anisms of evolutionary change. This fact is worth restating: you can
have evolution with or without selection. We've shown this to be true
with our population of Tadro3s. It's also worth pointing out that the
random genetic processes are always occurring and that they operate
independently from selection. There is a third mechanism influencing
how populations evolve, however: history (Figure 4.2).[6] The inde-
pendence of these three kinds of mechanisms was beautifully demonstrated by Professor Rich Lenski, of Michigan State University, in his

work first on bacteria and then on what he calls "digital organisms." Lenski showed that the genetic and phenotypic variation that exist in your population at any time constrain the population's future evolutionary possibilities. Any finite population can only evolve in certain directions, directions that are constrained by the underlying genetic coding of the phenotypes and the responses of individual organisms to the particular environment. Another way to think of this history effect is that selection can only select, as we saw in the Tadro3 generations 2 through 4, when individuals vary. Selection can only result in evolution-by-selection when genes code, at least in part, those individual variations.

For our population of Tadro3s, history covers a multitude of sins or, more accurately, assumptions. Our Tadro3s didn't first evolve from single-celled Tadro0.001s, so they don't have an explicit evolutionary history, but they do have an implicit one: the evolutionary history of the tunicate tadpoles after which Tadro3s are modeled. That tunicate history has bequeathed Tadro3s a notochord, a light-sensitive organ, a brain capable of linking sensor and undulating tail, and the genes that underwrite these phenotypes. History has also brought along environmental baggage—the water with a directional light source in which Tadro3 lives.

History's most important legacy for Tadro3s is the blood line, the initial genetic conditions—essentially the genes and the variants of those genes that we decided Tadros had. Because we were interested in structural stiffness of the notochord and, by extension, the tail in which the notochord sits, we coded for that property and placed its initial value in the middle of a scale of biological stiffnesses.

THE GENETICS OF STRUCTURAL STIFFNESS

Structural stiffness is a property that describes how a structure will change shape when an external force acts on it. Think about a cantilevered structure, like a flagpole hung horizontally. When you put a flag on the pole, it bends down slightly. Or, better yet, imagine that old Three Stooges movie short, *Flagpole Jitters*, in which Moe, Larry, and Shemp find themselves hypnotized and walking out on a flagpole high above the city streets. Through experiments with weights rather than clinging Stooges, engineers have figured out that the structural stiffness of a cantilevered beam like a horizontal flagpole is proportional to the beam's flexural stiffness and inversely proportional to the cube of the beam's length. This is one of those moments when, against the advice of my friends, I just can't help but use an equation to summarize all of this:

$$k = \frac{EI}{L^3}$$

To clarify, what this equation says is that structural stiffness, represented by the variable k (in units of Newtons per meter) is defined as the ratio of the flexural stiffness, the composite variable EI (in units of Newton square-meters), to the cube of length, L^3 (units of cubic meters). What's nifty about this equation is that you can see right away what matters. Want a stiffer beam? Increase the EI or decrease the L. A longer cantilever has a huge impact on k because of the cubing of the L. An equation like this also helps conjure up the genetics we're after.

We could have just said that Tadros have genes that code for k directly and then left it at that. However, research on biological stiffness suggests that all three variables, E, I, and L can change independently during development and evolution. The variable E (in measurement units of Newtons per square meter) by itself is called by a variety of names: "modulus," "elastic modulus," "complex modulus," "Young's modulus," or "Young's modulus of elasticity." Too many aliases! But wait—let me do my part for the witness protection of E. Because E is part of structural stiffness, k, and flexural stiffness, EI, and is caused by the kind and number of chemical bonds in the material you're dealing with, I prefer to call it "material stiffness."[7]

To allow for the likely possibility that selection might target the length of the tail for a variety of reasons, some of which may have to do with structural stiffness and some not, we created a genome that coded for L and E separately. By not coding for the variable I, we were holding that part of the geometry—and everything else about Tadro3 for that matter—constant. In the language of a geneticist, both L and E were quantitative characters, polygenes, multiple loci capable of producing smooth gradations in the phenotypes for which they code. All loci for L were located on a chromosome separate from the loci for E in order to allow for independent assortment. In other words, having quantitative traits means that the genome does not contain the simple on-off, wrinkled pea or smooth, kind of genes that we call "Mendelian." Each set of genes is, instead, a continuous number scale, capable, within a given window, of producing a range of E values different from a range of L values.

You can see the independent changes in the proportion of E and L genes in the bottom panel of Figure 4.1. Notice that as the proportion of L genes increases from generation 7 onward, the structural stiffness, k, in the middle panel, plunges. This is exactly what we'd expect from our equation for k, on previous page. That L^3 term in the denominator is increasing dramatically, and it is lowering k at the same time that the E term in the numerator is decreasing and also lowering k. Faced with this kind of genetic evolution, poor old structural stiffness doesn't stand a chance.

THE EVOLUTION OF STRUCTURAL STIFFNESS

Over the course of ten generations, the population's average value of the structural stiffness of the tail, k, plummets from above 5 to below 1 Nm^{-1}. We've seen what was happening genetically to cause the decreased value of structural stiffness. But these genetic changes don't speak to how selection—which judges individuals by their behavior, not by their genetics—was interacting with randomness and history. We still have two bothersome itches to scratch: (1) Why did the structural stiffness decrease under selection for enhanced feeding behavior when we predicted that it would increase? (2) Why does feeding behavior seem at times to be unrelated to the structural stiffness of the notochord?

I want to warn you right now about a tempting siren who begins singing on the rocks at about this point in a study. When, as was the case with Tadro3, your experiment produces a result that appears to be the exact opposite of what you predicted, the immediate emotional response is to be disappointed and self-flagellating. My students and I certainly were. When we graphed the data in Figure 4.1, we had to have a group counseling session immediately to air concerns and responses. In the lightning reaction round, we heard: What went wrong? Our experiment didn't work! These data suck! We stink as scientists! My line then, and I'm sticking to it now, is that if you design an experiment carefully, execute it well by tracking down mistakes as they occur and running controls, your data will always be great. Data just are. No matter what those data say about your predictions, they and the experiment that generated them stand on their own, with their total value determined by how well you measured what you set out to measure.

The negative emotional response, I reckon, comes from the fact that we all secretly think that we understand our experimental system well enough to know exactly how it will turn out. We are, from an emotional point of view, just running through the experiment to show other people what we've already figured out in our heads.[8] By the time we've made our prediction, a process that is really like running our own internal model cognitively, we have committed emotionally to a particular outcome. We aim to "prove" our point through demonstration.

Although disappointment and disillusionment in the face of unexpected results may be natural emotional responses—and ones that I share with my students—they run counter to the way that many but not all of us reason scientifically. Strictly speaking, we demonstrate that some testable concept is true through our repeated failure to show it to be false.[9] Although we can certainly demonstrate that something predictable happens every time we release our coffee mug from a height of two meters, no one has seen gravity.[10] Gravity is a concept for a kind of energy related to the masses of objects. The relationships that we see between objects on planets and between planets and stars in space is observable and consistent with our concept of gravity; hence, having failed repeatedly to disprove those consistent relationships between objects, most of us think that gravity is a fact. Because the failure to reject gives us confidence in the truth that we infer about gravity, we employ gravity, in turn, as an inferential tool to create new scenarios. We use the fact of gravity to infer the presence of the universe's unseen dark matter. If dark matter were shown to be nonexistent, that observation would indirectly refute the gravitational paradigm. In sum, to prove, we attempt to refute.[11]

Following our therapy session, which included a harnessed descent into the cave of refutational negativism, our Tadro team decided to look deeper into the data. We needed to figure out if, in fact, our data sucked (which is always a real possibility) or if the results were screaming in our faces about something really interesting that we just hadn't anticipated. I'll spare you the tedium involved in figuring out if your data suck: it's all the usual kinds of things about checking your transcriptions, lab notebooks, mathematical formulae, control experiments, and calibrations of instruments.[12] We came to the conclusion that we couldn't explain our results away with the reflexive "bad data by bad scientists." Something much more interesting was afoot.

If you look back at Figure 4.1, you'll notice that from generations 1 to 2, when selection was present, we had a big increase in the feeding behavior score that was correlated with a jump in structural stiffness. This pattern was as we predicted, and it allowed us to make the nifty overlay in Figure 4.3 in which the Tadro3 with the stiffest tail has the best feeding behavior and the Tadro with the most flexible tail has the worst. Also from generations 2 to 3, when selection wasn't present, we still see that feeding behavior and structural stiffness are correlated, changing together. All is well, even though stiffness is dropping a bit. No big surprise, given that random genetic changes can be dominant in the absence of selection, as we talked about earlier.

After generation 2 the system appears to run amok: from generations 3 to 4 and beyond, really, the one-to-one connection between behavior and stiffness disappears. Behavior drops or stays constant

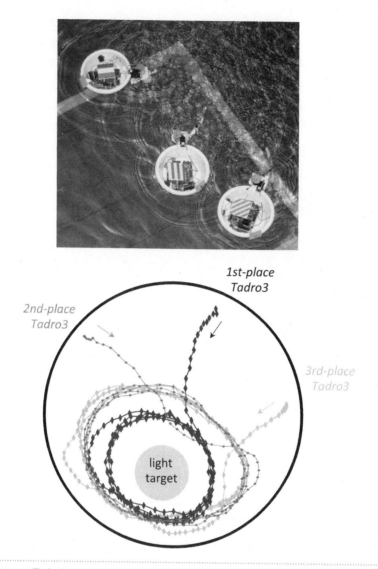

FIGURE 4.3. *Tadro3s compete for food.* In the top image the three Tadro3s jockey for position as they navigate to the light target, which serves as food. Note that the Tadro3 up top is taking a slightly different path than the other two. In the bottom diagram the paths of three Tadro3s competing in generation 1 are overlaid to show their differences in behavior. The first–place Tadro3 moves quickly to the light target, earning top scores as well for speed and the tight orbit it holds around the target. In contrast, the third-place Tadro meanders toward the target from its starting position and holds a much larger orbit around the target. For all three Tadro3s, the structural stiffnesses of their notochords positively correlated with their behavioral performance. This relationship between stiffness and behavior in generation 1 was reflected only in the evolutionary change in the population from generations 1 to 2 and 2 to 3 (see Figure 4.1).

while structural stiffness increases. Or the behavior score increases while structural stiffness drops, as in generations 4 to 5. In this case, in the absence of selection we know that we have only random genetic processes driving the evolutionary changes. Randomness challenges our assumptions. Because of the one-to-one relationship between behavior and stiffness from generations 1 to 3, we assumed that the structural stiffness of the axial skeleton was causally related to the feeding behavior. But faced with the evidence from later evolutionary changes, that relationship is either not true or it's only true some of the time. How can we tell?

SELECTION ON BEHAVIOR AND THE POPULATION'S EVOLUTIONARY RESPONSE

When complexity dims the light of interpretation, one way to navigate is to stop and examine your assumptions. In our case, the first assumption we tested was a fundamental one: when we selected for enhanced feeding behavior the population responded by evolving enhanced feeding behavior. This was what happened. Thank goodness! Selection was present in four generations: 1, 5, 6, and 9; in three of those cases (generations 1, 6, and 9), the ensuing generations showed higher average feeding scores than their parental generation (see also Figure 4.4). We took the data from each individual, not just the averages, from all four of those generations with selection and statistically tested the mean response to selection. The statistical test confirmed what we see by eye: on average, selection evolves enhanced feeding behavior. Keep in mind that even when selection is acting, randomness will almost always deflect, to lesser or greater degree, the population's evolutionary trajectory that selection proposes (see Figure 4.2).

We can learn something new (for us) and important by examining the only time, from generation 5 to 6, when selection didn't evolve improved feeding behavior. If you look at the change in genes that results from selection (bottom panel, Figure 4.1), you can see a drop in tail length, L, accompanied by a jump in material stiffness, E. Going back to our handy-dandy equation (I just knew it would be useful) for structural stiffness, k, we know that because of the magnifying effects of the L^3 term in the denominator, the population's average k must be higher in generation 6 than in generation 5. And, indeed, we see that jump in average k in the structural stiffness plot just above the genes plot. This connection of genes and increased structural stiffness rules out the random genetic deflection idea in this case as the primary cause of the evolution of the feeding behavior. To explain this fully, though, we have to go on a little journey. Fasten your safety harness, please.

What the disconnect between selection in generation 5 and feeding behavior in generation 6 suggests is that we need to test this assumption: feeding behavior is causally connected to structural stiffness of the notochord. If this assumption were always true, then we'd expect to see behavior improve or decline in concert with increases or decreases, respectively, in the notochord's structural stiffness. From our previous discussion, we know that behavior and stiffness don't show any regular pattern that might lead us to believe that they were causally connected. However, several other patterns are possible. First, it could be that changes in the notochords' structural stiffness are correlated not with the overall feeding behavior but rather with some of feeding's sub-behaviors: swimming speed, body wobble, average distance from the food, and time to find the food. Second, those sub-behaviors might not be correlated with structural stiffness but rather with the stiffnesses' subcharacters: material stiffness, E, and length of the tail, L.

We ran a series of statistical tests to look at the patterns of correlation of the stiffness variables on one hand—structural stiffness, k, material stiffness, E, and length of the tail, L—and the behavioral variables on the other—swimming speed, V, body wobble, W, distance from food, D, and time to find the food, T. As separate independent variables in a linear regression, k, E, or L all predict about 20 percent of the variation in V and W but predict none of the variation in D and T.[13] Moreover, k and E are positively correlated with V and

W, and *L* is negatively correlated with *V* and *W*. Thus, we've got a clear relationship between structural stiffness of the notochord and swimming speed and body wobble, two of the four components of the feeding behavior score.

What's strange about this pattern of correlations is that swimming speed and body wobble are positively correlated. Recall that we had predetermined that the fitness function would judge increased speed as "good" and increased body wobble, distance to the food, and time to the food as "bad." Thus, the fitness function should, when selection is strong enough to create differential reproduction, create this specific pattern of correlations among the sub-behaviors (Figure 4.4). Instead, the evolutionary pattern that we always got was the one we just identified with statistics on individuals across generations: speed and wobble always increased together under selection.

Surprise! When we select for improved feeding behavior we actually make the evolutionary fitness of the Tadro3s better and worse at the same time. Tadros swim faster (and that's good for an individual's fitness score) but they wobble more (and that's bad for the fitness score). To explore what was going on with this odd couple, we swam our Tadros with all their different tails in a simple swimming trial— no competition, no evolution. Just swim straight down the tank. We videotaped the Tadro3s and measured their speed and body wobble. Under these conditions speed and wobble were not correlated. In other words, in the "wild," competing with other Tadro3s to swim and feed, speed and wobble were, for some reason, functionally connected. But in the "lab," swimming by themselves in a straight line, Tadro3s didn't show this functional connection.

All is revealed when we look closely at this sub-behavior that I've been calling body wobble. You may recall from the previous chapter that we said that wobble was a measure of how unsmooth the swimming path of the Tadro was. If you ever read Arthur Ransome's *Swallows and Amazons* series as a kid, you'll know that you can spot a novice helmsman on a small sailboat by the wiggles in the boat's wake. Those wiggles, which come from an unsteady hand on the

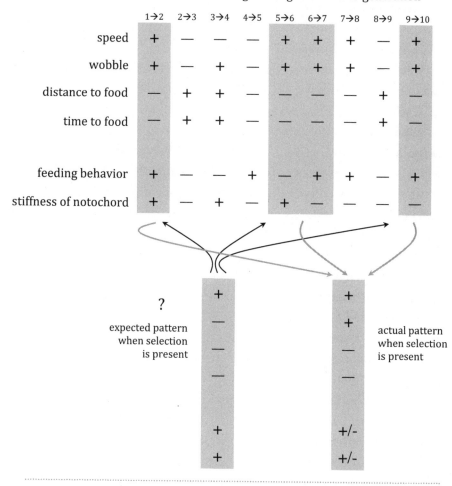

direction of change from generation to generation

FIGURE 4.4. *For the sub-behaviors that make up feeding behavior, a consistent pattern of evolutionary change occurs when selection is present in the population of Tadro3s.* For swimming speed, body wobble, average distance to the food, and time to find the food, selection, indicated by the gray background, always increases speed and wobble while decreasing distance and time. The actual pattern (bottom, right) differs from the expected pattern (bottom, left), which is based on the fitness function assigning a penalty to increased body wobble. The consistent correlation of speed and wobble under selection indicates that the two are functionally linked when Tadro3s are competing for food. When we examine the stiffness of the notochords across all generations, we find that it is statistically and positively related to swimming speed and body wobble. Stiffness is thus indirectly, not directly, correlated with feeding behavior through speed and wobble.

tiller, waste energy by rotating the boat back and forth, slowing down the boat's forward progression. We took Ransome's wiggles and turned them into wobbles. From videotape records, we measured the change in heading of each Tadro3 every second during the three-minute trial. We then calculated, over every two seconds, how rapidly the change in heading was occurring. If you are a sailor who navigated high school physics successfully, you may recognize this as a measure of angular acceleration in yaw. We then took all of the angular accelerations for the whole trial and calculated their standard deviation, which is a measure of how variable the angular acceleration was. This gave us a single number called wobble (in units of radians second^{-2}).

Check it out: wobble not only measures the rate at which the Tadro is wiggling and losing energy, what we call recoil from the flapping tail, but it also measures the presence of high-speed maneuvers. Think of it this way: a quick turn is a big, quick wiggle. We figured this out by running some digital simulations in which we would make a swimming Tadro perform a big turn quickly. We measured the resulting increase in wobble. It turns out that frequent turning maneuvers produce a huge amount of wobble, much more than the energy-wasting recoil.

We realized that wobble in the wild was picking up on all of Tadro3s quick turning maneuvers that weren't present in the lab's straight swimming. Cool! We had accidentally measured something different and even more interesting than what we'd intended. Now it made sense why swimming speed and wobble were positively correlated in the wild: if you are swimming faster you can turn faster. Plain and simple. Every driver and sailor knows this and slows down coming into a turn in order to maneuver smoothly or, instead, maintains their headway to make a break-neck turn.

In Tadro3s we now see that body wobble is functionally dependent on swimming speed. This gets us into a really interesting area in terms of evolutionary biology. Any trait that has a genetic basis and whose phenotypic expression is influenced by a different gene is said to be "epistatic."[14] Epistasis is a very important genetic phenomenon and can occur in a variety of ways that biologists are still uncovering. With Tadro3s, though, the interaction is not conducted directly at the genetic level; we even set up the genes to make sure that they didn't interact at the level of the genome. Instead, here we see phenotypic epistasis, a physical interaction of the sub-behavior phenotypes that occurs because the phenotypes share a single body. We established the genetic basis of speed and wobble through their connection to structural stiffness of the notochord, which is directly coded by the Tadro3s' quantitative genes.

Now that our wobbly journey through epistatic seas is over, we can stand on *terra firma* when we explain the evolutionary change, under selection, from generation 5 to 6 (see Figure 4.1, again!). We had observed that this was the only time, out of four evolution-by-selection events, when selection made swimming behavior worse. We had established the fact, by looking at the changes in the proportion of genes, that random genetic effects couldn't explain the evolutionary change. What was left to analyze was the functional relationship between the notochord's structural stiffness, which is coded genetically, and the feeding behavior, which is judged by selection. One of the sub-behaviors, body wobble, turned out to be beneficial to improved feeding behavior, instead of detrimental, as originally thought. In addition, body wobble is functionally linked to swimming behavior.

These two final and unexpected facts about body wobble explain the also-unexpected degradation of feeding behavior from generations 5 to 6. Because the fitness function rewards increased swimming speed while penalizing increased wobble, the composite score of feeding behavior drops. Keep in mind that we measured feeding behavior using the same relationships that we used in the fitness function.

Knowing what we know now, this was a mistake. We should've rewarded increased wobble and called it something more appropriate, like "agility." Unfortunately, we can't go back and change evolutionary history without completely redoing the experiments (more on why in the following pages). We calculated the fitness function every generation in order to make the next generation. But we can recalculate the feeding behavior score with increased wobble rewarded, not penalized.

When increased wobble is rewarded, we get a slightly and importantly different picture of the evolution of the composite behavior that we call feeding of Tadro3s (Figure 4.5). First, the averages of the population's feeding behavior now undergo more change, moving both higher and lower than their means under the old fitness-based measurement scheme. Second, one of the generation-to-generation evolutionary changes is different: from generations 5 to 6, under selection, average feeding behavior now improves rather than degrades. In other words, armed with our new understanding of wobble as a positive metric of rapid maneuvers, feeding behavior actually did improve under selection!

This seems to be a most ingenious paradox. We are saying that feeding behavior declined from generations 5 to 6, but really, it improved. What? Allow me to sum up. Before the evolution of Tadro3s started, we created a fitness function that we thought was selecting for improved feeding behavior. This fitness function rewarded increases in swimming speed and penalized increases in body wobble, time to find the food, and distance from the food. In each generation, if individual Tadro3s varied enough in their feeding for the mating algorithm to create differential reproduction, then selection was both present and active as an evolutionary mechanism.

In only one case, generations 5 to 6, was selection present when average feeding behavior declined. We had measured the feeding behavior of individuals using the same four sub-behaviors and their

FIGURE 4.5. (facing page) *The evolution of feeding behavior revisited.* When increases in wobble, which correspond to quick turning maneuvers, are rewarded in the new metric (black line labeled as "NEW") rather than punished in the old (gray line labeled as "OLD"), feeding behavior has the same general pattern as before but with two important exceptions. First, the new average feeding behaviors are both higher and lower than the old averages. Second, the evolutionary transition from generations 5 to 6 is positive with the new metric, so that in all cases when selection is present, feeding behavior improves.

phenotypes

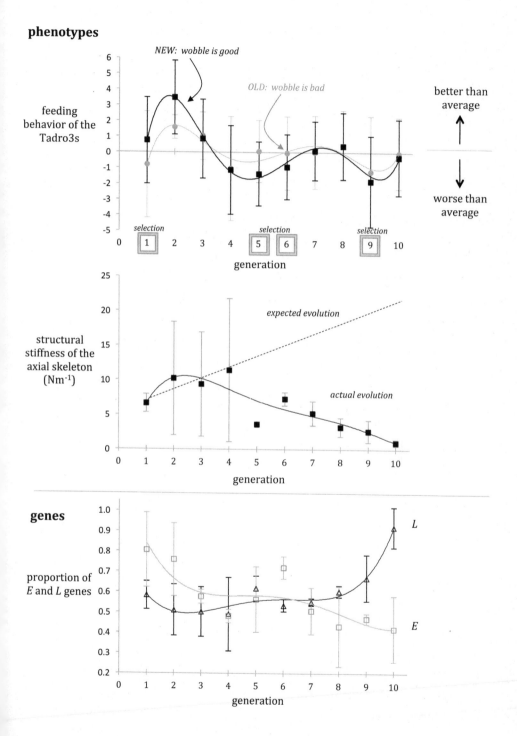

feeding behavior of the Tadro3s

NEW: wobble is good

OLD: wobble is bad

better than average

worse than average

selection

generation

structural stiffness of the axial skeleton (Nm⁻¹)

expected evolution

actual evolution

generation

genes

proportion of E and L genes

L

E

generation

goodness or badness as used by the fitness function. All that we changed in making the behavioral metric was to compare individuals not just in a generation but instead across all generations as well as to scale individual differences relative to the variance exhibited by all Tadro3s across all generations. Under this original fitness-based metric, we found that one apparent anomaly in feeding behavior.

We ruled out random genetic change as the cause of the decrease in feeding behavior from generations 5 to 6 because the proportion of genes changed to increase structural stiffness of the tail, and structural stiffness of the tail is positively linked to swimming speed. That left us to reconsider the four sub-behaviors and their functional interactions. We discovered that in the competitive arena, increased wobble wasn't a sign of energy inefficiency but rather agile turning maneuvers when facilitated by high swimming speed. Hence, our original idea about what constitutes good and bad feeding behavior was just plain wrong.

When we altered our after-the-fact behavioral metric, we found that feeding behavior always, in fact, improved under selection. The paradox arose because the selection was actually penalizing increased wobble when, as we know now, it should have been rewarding it. As I mentioned earlier, we would love to revise the fitness function and repeat the experiment. We'd expect that selection, when present, would be even stronger, and that the jumps might be greater. But we have one big problem with repeating the work: evolution of physically embodied Evolvabots takes loads of time and buckets of money. This is one reason you try to be as careful as possible in the design phase (see Chapter 3)!

WHAT HAVE TADRO3S TAUGHT US?

We thought that our Evolvabot design, carefully laid out as a series of simplified representations of nature in the previous chapter, would produce a simple evolutionary pattern. We could not have been wronger.[15] We evolved two phenotypes, material stiffness, E, and length of the tail, L, that together are responsible for the structural stiffness, k, of the notochord. These traits were coded as quantitative genes housed in a diploid genome. The possessors of the tails whose phenotype was dictated by the genes competed for food. Selection, codified by our fitness function, rewarded individuals in a particular generation who behaved better in terms of increased swimming speed, decreased body wobble, decreased average distance to the food, and decreased time to find the food. For reproduction, haploid gametes were mutated and combined in a simple random mating scheme to produce the instructions for the notochords of the next generation.

Our first surprise came when we saw, after ten generations of a constant selection pressure, that the evolutionary changes in the population's feeding behavior, tail stiffness, and gene proportions were anything but constant. Why would a constant selection pressure produce different results each generation? Part of the answer is that in each generation the other agents in the world have changed, and their different evolved behaviors alter the competitive landscape. Another part of the answer is that genetic variability contracts and expands over generational time, changing the phenotypic options available for selection to judge.

Our next surprise came when we realized that selection was only operating to produce differential reproduction in four of the ten generations. This meant that in the generations without differential reproduction the evolutionary changes in phenotype and genotype were happening because of purely random effects. In particular, mutation and genetic drift caused mutational differences and individual genomic differences to combine into relatively large effects when selection was not present.

Our final surprise came as we probed the causal connections between the structural stiffness of the notochord and feeding behavior. Feeding behavior was measured by the same sub-behaviors that we put into our fitness function—swimming speed, body wobble, average distance to the food, and time to find the food. When we correlated these sub-behaviors with structural stiffness of the notochord, we found that swimming speed and body wobble were positively and significantly correlated with structural stiffness, k, material stiffness, E, and length of the tail, L. Time and distance to the food were not. This meant that when time and distance were undergoing greater evolutionary changes than were speed and wobble, structural stiffness could be decoupled from feeding behavior. This situation was complicated by the fact that speed and wobble are positively correlated in terms of function but are negatively correlated in the fitness function. Hence, in terms of fitness, their effects would tend to cancel.

REFUTING A HYPOTHESIS

Now that we are confident in our understanding of the mechanisms and interconnections driving the evolution of Tadro3s, we can be confident in doing what we meant to do all along: not just create an evolutionary simulation but test a hypothesis about the biological system that the simulation represents. We proposed the hypothesis that natural selection for enhanced feeding behavior drove the evolution of vertebrae in early vertebrates. From this hypothesis we came up with a primary prediction that we tested: selection for enhanced feeding behavior will cause the population of Tadro3s to evolve stiffer tails.

Our data refute this prediction (see Figure 4.1). Hence, the hypothesis from which it came is also refuted. We found just the opposite of our expectation in some cases—selection for enhanced feeding behavior caused the population of Tadro3s to evolve more flexible tails (see Figure 4.1, generational change in stiffness following selection). If we buy our own argument that Tadro3s and their water world represent important aspects of the earliest vertebrates, we have to conclude that selection on feeding behavior was unlikely to have been the primary driver of the evolution of vertebrae. In science this kind of failure is called progress.

But if not feeding, then what drove the evolution of vertebrae? The positive relationship that we've shown between tail stiffness and swimming speed + turning maneuvers offer one alternative hypothesis: selection for speed and maneuverability alone drove the evolution of vertebrae. The problem is that in order for selection to work on locomotor abilities alone, it can't be simultaneously working on other things like feeding behavior. If it does, we get these oscillating patterns of change that we saw in this experiment, when stiffness is sometimes correlated with feeding behavior and sometimes not.

This kind of complex response to selection, by the way, is very realistic in living fishes, as David Reznik of the University of California and his colleagues have shown.[16] Based on extensive review of the literature on responses of fish populations to selection, they contend that a behavior like acceleration performance is influenced by a network of traits, all of which, because of their genetic properties and functional connections to other behaviors, can be under countervailing selection pressures at the same time. Dominant selection pressures on a population can also vary in the wild, Reznik and his colleagues have shown, as predators move in and out of small pools containing mating subpopulations of Trinidadian guppies.

Predators appear to create strong selection pressures in other species of fish as well. Dohlf Schluter and his colleagues at the University of British Columbia have shown that moving threespine sticklebacks from ocean habitats to freshwater lakes—migrations that occur naturally—appears to release the immigrant population from predation pressures found in marine environments.[17] Without predation from other vertebrates, the sticklebacks respond genetically and, over generational time, grow faster and produce less body armor. Moreover, Richard Blob, of Clemson University, and his colleagues suggest that predation may have provided the selection pressure for stream gobies to evolve a remarkable ability to scale waterfalls in Hawaii.[18]

A different kind of interpretation of our results is that we didn't really test a hypothesis about vertebrae because we had used the stiffness of the notochord as a proxy for the number of vertebrae. Even though this relationship is mechanically based, with stiffness and vertebrae being positively related, what if something else about vertebrae mattered? Perhaps the presence of vertebrae increases stiffness *and* changes the curvature of the tail, its shape, and the way it acts as a propeller?

Another valid criticism of our work with Tadro3 is that its brain was just too simple to adequately model even a simple system like the tunicate tadpole larva. But, we counter, the point is to create the simplest system possible because even the simplest autonomous agents produce complex behaviors and complex evolutionary patterns. You test your hypothesis using the simplified model. Then, based on the results of the test, you interpret what happened; you work to understand the mechanisms operating at multiple levels in your system. Interpretation, as we've seen, is darned tricky, even when you operate under the KISS principle.

What we've learned from Tadro3 is that neither the selection pressure, the design of Tadro3 itself, nor an interaction of the selection environment and the Tadro3 agent explain why vertebrae are likely to have evolved. We haven't solved the puzzle!

So our next step is to think about how to make Tadro and the selection pressure a bit more complex. Thinking about our design principles from Chapter 3, we have to make sure that we understand the biology well enough so that we can understand what it is that we want in our Tadro4 and its world. We've just set up predation as a great candidate for an additional selection pressure. What we need to know much more about, though, is this tricky business of brains and behavioral complexity. At the very least Tadro4 will have to be a smart prey that is able to eat and, at the same time, avoid being eaten.

chapter 5

THE LIFE OF
THE EMBODIED MIND

SOMETHING STRANGE JUST HAPPENED IN THE LAST chapter. Did you notice? When we applied selection pressure on our Tadro3s they responded by evolving next-generations of smarter Tadro3s with better feeding behavior than their parents had—in a real sense, they got smarter. But when our population of Tadro3s became smarter, they did so by evolving their bodies, not their brains.

How in the artificial-water-world can Tadros, or any robot for that matter, become smarter? And even if they can gain intelligence, how can intelligence be "in the body" rather than "in the brain"? Isn't the brain the seat of intelligence? And by the way, as long we're inquiring, where is the brain of a Tadro3 anyway, and what is it doing?

We need to tackle these questions because they help put Tadro3, which you met in the last chapter, and Tadro4, which you'll meet later in this chapter, into the broader context of intelligent machines. Although I'm not claiming that either Evolvabot is going to win a merit scholarship to attend Vassar, I will claim that Tadros—by virtue of being goal directed, autonomous, and physically embodied—have intelligence. Hang on: the ride is going to be bumpy!

THE SEARCH FOR INTELLIGENT (ARTIFICIAL) LIFE

What we've got here, thanks to Tadro3 (there, I'll blame the robot), is not a failure to communicate but rather an opportunity to lay bare some of our conceptual problems. Most humans would argue, for example, that Tadro3s are not intelligent. Yet clearly, the autonomous, self-propelled Tadro3 has something. Let's call it skill: the ability to detect light, move toward it, then hang around it. Yes, say the humans, but moths possess the same skill, and we know that moths aren't intelligent. We do? What do you mean by "intelligent"? Intelligence is not just some simple skill, they say, like detecting and finding light; that's more like a reflex. Instead, the argument goes, intelligence involves the skill of thinking, using our minds in the special, linguistic way that only humans do. Thus, many humans speak of "human-like intelligence" as the sine qua non of intelligence.

Where does that leave the rest of the world? Are nonhuman primates intelligent? Is your dog intelligent? How 'bout your African Gray parrot? If you answer either "yes" or "no," then answer this: how do you know?[1] One answer, given by Alan Turing, is that you know by interacting.[2] He asks, "Can machines think?" In terms we've been using, Turing comes up with a way to answer this question—with the assumption that thinking is proof of intelligence—by creating an environment that includes at least two agents, each being a part of the other's environment.

For Turing, the interaction between agents and their environments is conducted using language: agents, hidden from each other, converse via keyboards. The brilliance of this conversational environment is that it is dynamic and far-reaching: linguistic communication in real time is a back and forth that can be about anything. If the machine, in practice an artificial intelligence (AI) in the form of a computer program and its hardware, can fool its human interactant into thinking that the human is conversing with another human, then the AI is said to have "passed" what we have come to call the "Turing Test." The Loebner Prize Competition is a Turing Test held every year as an international competition.[3] A bronze medal, along with a cash prize, is awarded to the AI that fools the most human judges. A gold medal awaits the AI that is indistinguishable from a human. We are still waiting for an AI to claim that prize.

An even tougher test of human-level intelligence is what Stevan Harnad calls "the total Turing Test."[4] In the total Turing Test (T³) the AI has to be embodied and physically present in an environment shared with the human interrogators. In other words, the AI has to be an embodied robot, and human-level intelligence is only achievable with a body and a brain. The embodied robot must be able to physically perform, in all ways, as an indistinguishable member of the group of organic agents to pass the T³. You can see that the T³ is a tall order, especially if you think of humans and the human interactional environment: language, movement, and physical appearance—all have to be on the mark, like a teenager struggling to fit in. In the human arena robots are not even close to competing at the bronze-level equivalent of the T³. At the moment the T³ for humans is the stuff of science fiction, like the replicants in the movie *Blade Runner*.[5]

In opposition to the interaction-based Turing Test, John Searle takes a different approach to the search for intelligence.[6] He looks for systems that understand what they are thinking about. For example, you know that you are thinking right now because you can use the symbols of written or spoken language to talk to yourself or to others about your thinking. You understand that you are "expressing" yourself. Your subjective first-person experience as the agent doing the expressing allows you to know that the word symbols you manipulate in your speech or writing contain meaning. You have the ability to analyze your own mental states, and by so doing, you are aware of your own intelligence and the processes that underwrite that intelligence. You can verify that your linguistic symbols have meaning to you.

This ability to be aware of ourselves analyzing ourselves is why human scientists get excited when a dolphin recognizes itself in a mirror.[7] Diana Reiss and Lori Marino, the researchers who've done this work, show behavioral evidence of the dolphin's self-awareness. The dolphin looks in the mirror, sees a spot of investigator-applied zinc oxide on the dolphin in the mirror, and then proceeds to spend time turning its body to examine the body of the dolphin-in-the-mirror for other blemishes. Reiss and Marino interpret this behavior as showing that the dolphin understands that the image in the mirror is representing "self" and not "other." Pretty cool.

Distinguishing between yourself-as-an-agent and others-as-agents is the basis of inferring that other agents may be intelligent. We have the ability to make this distinction, and we use that ability to infer that other conscious, human agents possess the same ability. We can report those subjective experiences to others using language: I know that I'm intelligent; I know that you are a human agent like me; therefore, I infer that you are intelligent like me.

When we search for intelligent life, we combine the approaches of Searle and Turing. First, I understand that I'm intelligent because I'm the "I" experiencing my intelligence (Searle's criterion). Second, I'm guessing that you are intelligent because when we interact you behave in ways that make me think that the only way we can be having an interaction like this is if you have an intelligence very much like my own (Turing's test).

Most people, when asked by Daniel Wegner and his colleagues at the Mental Control Laboratory at Harvard, say that other human beings have features that we associate with an intelligent mind: consciousness, personality, feelings, emotions, rights, responsibilities, self-control, planning, thought, and recognition of emotion in others.[8] A surprise is that these same humans perceive that some of these mind-like features are possessed, to varying degrees, by entities that include the nonliving, such as God and robots. If it's fair for me to perform the sleight of hand that equates mind-likeness and intelligence, then we twenty-first-century humans readily perceive intelligence all over the place. Perception, however, is not necessarily reality.

TADRO'S KNOW-HOW

When Adam Lammert and I showed my colleague Ken Livingston the first working Tadro, Tadro1, we were excited and a bit nervous. Livingston, professor of psychology and one of the founders of the Cognitive Science Program at Vassar,[9] had served for both of us as a mentor in the ways of embodied robotics and artificial intelligence. When he saw Tadro swimming around in a big sink in the lab, following the beam of a flashlight we were moving around, he grinned and said, "Tadros are a piece of embodied intelligence." Would you agree?[10] Adam and I did. Here's why.

Putting Turing's hat back on, let's think about what we were doing with Tadro. We put a Tadro in the sink, turned off the lights, and then turned on a flashlight. The Tadro, which had been aimlessly swimming around the tank, changed course with what looked like purpose and curved in a right-handed loop toward us, bumping the wall of the tank, turning around to the left, and then heading back in our direction. We then played a trick: lights off. Because we've put green and red navigation lights on Tadro, we could see Tadro in the dark as it changed the curvature of its heading, moving now in a left-handed arc along the wall of the tank. We snuck around to where Tadro was headed, and surprise! We turned on the flashlight directly over Tadro's head. Tadro's response was immediate: a quick turn to the right, moving off the wall, and heading back into the darkness.

Okay, so is this the most fun we've ever had? Nope, but it beats washing the dishes. When you play around with Tadro, you experience a sense that you have to learn, through your interactions, about what's not predictable and what is. You can't predict exactly what Tadro will do, how much it will turn, where it will hit the wall of the sink. At the same time, you learn very quickly that in response to light on its single eyespot, Tadro turns to the right. When that eyespot is in darkness, Tadro turns to the left. You even figure out that you can interact with Tadro in such a way as to get it to swim straight for just a bit when you find just the right light intensity that is midway between full dark and full light.

Now let's take off Turing's hat and put on Searle's. We immediately turn on the lights and pick up the dang Tadro. What is this thing? What's inside? We look inside the plastic bowl that serves as Tadro's hull and see a small, black, rectangular box with what looks like a big, engorged tick sticking on it (sorry, I live in upstate New York, one of the world's hotspots for blood-sucking ticks and the diseases that they spread; I see them everywhere . . .). That "tick" is actually a capacitor, a common element of electronic circuits, and it is interspersed with other bug-sized bits of like-minded paraphernalia: rectangular silver-legged spiders (integrated circuits), columns of red and green "ants" (indicator lights), and the long tracks of tiny potholes left by centipedes (input and output connections for wire). This palm-size block of electronics is a microcontroller,[11] a fully functioning computer (a central processing unit, or CPU) with its own power supply, memory, and systems to operate motors and sensors.

Is this microcontroller a brain? Doesn't look like it. It's a computer with the ability to interact with motors and sensors. You can program the microcontroller to tell the tail motor which way to turn depending on the light intensity hitting the photoresistor that serves as the eyespot. The program is not the brain, either. It is written in a programming language called Interactive C, which was created especially for controlling mobile robots.[12] We can call up the original Interactive C program that Adam Lammert wrote for Tadro2 and see for ourselves that there appears to be nothing brain-like about it (Figure 5.1): a bunch of words and type-written symbols, a regularity of symbol patterning that indicates a syntax, and words like "if" and "else," which if used in the same way as those words are in English, may indicate something about the program making decisions. Even if that naïve description sounds brain-like because of its references to language, keep in mind that Searle would argue that the Tadro program is not, in and of itself, intelligent; the program doesn't know what it is doing. It just is a deaf, dumb, blind kid telling the hardware how to play pinball with electrons.

```
#use "servo.icb"
//----------------------definitions----------------------------
#define BETA_MIN 0.0
#define BETA_MAX 120.0
#define SENSOR_MAX 45.0
#define SENSOR_MIN 5.0
#define SERVO_TIME 50L
#define SERVO_RANGE (MAX_SERVO_WAVETIME-MIN_SERVO_WAVETIME)
#define rexcursion 3.14159
#define dexcursion 180.0
#define REMOTE_BUTTON 80
//----------------------declarations----------------------------
float Beta;
int Sensor_int;
float Sensor_float;
float Seconds = seconds();
int Whole_Seconds = (int)Seconds;
int Num_Seconds = 10;
int Print_Period = Whole_Seconds%Num_Seconds;
int Print_Track = Print_Period;
//----------------------program---------------------------------
void main()
{
    msleep(500L);
    servo_on();
    while(1)
        {Seconds = seconds();
         Whole_Seconds = (int)Seconds;
         Print_Period = Whole_Seconds%Num_Seconds;
         Sensor_int = analog(3);
         Sensor_float = (float)Sensor_int;
         if(Sensor_float > SENSOR_MAX)
            Sensor_float = SENSOR_MAX;
         if(Sensor_float < SENSOR_MIN)
            Sensor_float = SENSOR_MIN;
         Sensor_int = (int)Sensor_float;
         Sensor_float = 50.0 - Sensor_float;
         Beta = ((BETA_MAX/SENSOR_MAX)*Sensor_float) +
                ((dexcursion-BETA_MAX)/2.0) + 15.0;
         servo_deg(Beta);
         if((Print_Period)==0 && (Print_Period) != Print_Track)
            {if(Sensor_int>9)
                printf("%d", Sensor_int);
             else
                printf("0%d",Sensor_int);}
     Print_Track = Print_Period;}}
```

sensory input creates motor output

FIGURE 5.1. *Making Tadro go.* This is the complete Tadro2 program, written in computer language called "Interactive C" by Adam Lammert for his senior thesis in cognitive science at Vassar College. The program ran on a HandyBoard microcontroller, taking input from the single sensor, a photoresistor acting as an eyespot, and turning it into a value for the variable, "beta," that told the always-flapping tail which way to turn. This sensor-motor interaction, shown in the gray box, is always changing because the new turning command alters the heading of the Tadro that, in turn, alters the light hitting the sensor. Thus, the whole system—program, micro-controller, sensor, motor, body, and environment—can be thought of as continually calculating an answer to this question: what's the angle of my tail?

See the problem here? If we define intelligence by what Tadro does, then it clearly has skill, the know-how to detect and follow a light source. Because Tadro's light-following ability depends on its propulsion, maneuverability, and the sensitivity of its photoresistor, its body is clearly important. You can get a sense for the importance of having a body to help you think next time you put together a difficult jigsaw puzzle: you simply can't solve the puzzle unless you allow yourself to pick pieces up, rotate them in different directions, and try to align and engage the pieces.

If you don't believe me, try this: have a friend spread out pieces of a jigsaw puzzle on the table. *Rule 1:* You are not allowed to touch the pieces. *Rule 2:* You are not allowed to move from where you sit or stand. *Rule 3:* Using only your voice (not gestures or written instructions), tell your friend how to assemble the puzzle, piece by piece. Note that you can't just say, "Put the puzzle together." No how. You've got to give low-level instructions like, "Take the piece right in front of you and move it next to the piece right over there." *Rule 4:* All your friend can do is follow your rules. These rules turn out to be what we call motor commands when we talk about neural circuits. You'll soon be impressed—unless you have a very simple puzzle—with just how much intelligence depends on your movements and your physical manipulation of the world. That movement-based intelligence begins with what Alva Noë, associate professor at the Institute for Cognitive and Brain Sciences at the University of California at Berkeley, calls "enactive perception."[13] Enactive perception in robots combining active vision and feature selection helps simplify vision-based behavior, as shown in experiments by Dario Floreano, director of the Laboratory of Intelligent Systems at the École Polytechnique Fédérale de Lausanne in Switzerland and one of the founders of the field of evolutionary robotics.[14]

Adam and I give Tadro credit for having the know-how of enactive perception: wandering around, exploring its space, detecting a light gradient (if it's there), moving toward the source of the light, and orbiting around that source. If you had to tell your puzzle-helping friend to do the same, step by literal step, I bet you'd learn to respect our piece of embodied intelligence!

THE EMBODIED-BRAIN OF TADRO3

With the hats of both Turing and Searle off, let's baldly go where no one has gone before: into Tadro's embodied-brain. I say "embodied-brain" as a single-word construct here because I want to reference a shift in perspective for neuroscientists promoted by Professor Barry Trimmer, neuroscientist and director of Tuft University's Biomimetic Devices Laboratory. I visited his lab recently, and as we were discussing how animals create behavior, he said, "Every brain has a body." Sounds straightforward. But wait—that seemingly self-evident phrase, familiar to many within the fields of philosophy of mind,[15] ecological psychology,[16] grounded cognition,[17] and embodied artificial intelligence,[18] rings wrong-headed to neuroscientists who've specialized in the brain's molecular channels, neurotransmitter systems, control circuits, or functional regionalization. Why?

Most of us have been trained to think of the brain as the control center, the place on the anatomical map where all of the sensor inputs are read and discussed. We know that the brain is "in control of behavior" because damage to the brain alters our thinking: damage to Phineas Gage's frontal lobes compromised his ability to process emotions and make rational decisions.[19] We've seen cool functional-MRI videos of Oliver Sack's brain responding differently to music by Bach and Beethoven, in concert with his reported subjective experience.[20] After much thinking, our subjective, first-person experience of being an autonomous agent tells us that the control center creates a plan that is sent out to the soldiers in the field, the muscles that put the plan into action.

As neuroscientist Joaquin Fuster of the UCLA Neuropsychiatric Institute more formally states, "All forms of adaptive behavior require the processing of streams of sensory information and their transduction into series of goal-directed actions."[21] Fuster reviews experimental work that shows how goal-directed plans activate the prefrontal and premotor regions of the brain. In this view, planning is a central, if not *the* central, function of our brains as thinking machines (Figure 5.2).

In a move that Oz might be tempted to call the "reverse Scarecrow," Fuster takes care to give the brain a body. The prefrontal and premotor regions of the action-planning brain are part of what he calls the "perception-action cycle," which is "the circular flow of information from the environment to sensory structures, to motor structures, back again to the environment, to sensory structures, and so on, during the processing of goal-directed behavior." Pushing this point further, Trimmer, whose team specializes in designing soft-bodied robots to test ideas about how caterpillars move and modulate their behavior, said at our meeting, "The body is doing the computational work of interacting with the environment." But what is the nature of the "computational work" being done?

The body of Tadro3 "computes" everything that the microcontroller running the Interactive C program (see Figure 5.1) doesn't: all the really difficult physics. By virtue of being in the real world, interacting with real water, Tadro3 automatically solves the intensely complex dynamics of a flexible propeller transducing an oscillating uniaxial bending couple into a propagating bending moment that flexes the tail, which, in turn, is also loaded hydromechanically in a time-varying manner as its relative motion in the water changes. In response to the tail's coupled internal and external force computations, the body, to which the tail is attached, undergoes the yaw wobbles—recoil and turning maneuvers—that we talked about in Chapter 4. Coupled computations that allow elastic and fluid forces to interact have been elegantly simulated by Eric Tytell, of Tufts University, and his colleagues at the University of Maryland using the "immersed boundary method" for a steadily swimming lamprey.[22]

But wait. Order before midnight and your Tadro3 comes with free motor and sensory computations. Tadro's rotational and translational motion has angular and linear components of both velocity and acceleration that interact to produce the overall motion of the Tadro ac-

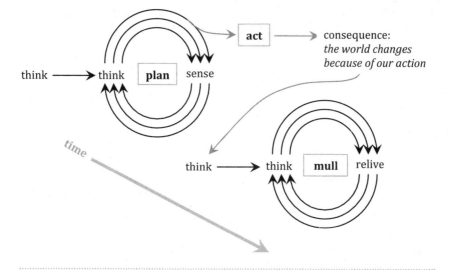

FIGURE 5.2. *The neurocentric view: thinking is planning.* Planning is something we do "in our heads," in our brain, with input from our senses, to create actions. This view is consistent with our subjective experience, coupled with information from neuroscientific studies of brain activity correlated with what we are thinking. The neurocentric view dominates in nearly every field concerned with human thought, language, and behavior.

cording to Newton's laws of motion. As the Tadro3 wobbles and winds its way through the water world, it presents its attached photoresistor, acting as an eyespot, to a gradient of light. As the light intensity at any place on the water's surface changes as the Tadro moves, the photoresistor continuously recomputes light intensity as a change in voltage by virtue of being part of a little electric circuit that works via Ohm's law.

However impressive Tadro3 might be as a student of physics, you may be objecting to the notion that Tadro3 is "computing" or "solving" anything with its body. "Computing" has a formal definition that goes back to Turing. We talk about "Turing-computable algorithms" as those procedures that can be solved, ultimately, with the simple deterministic rules a digital machine puts into play. Meanwhile, "solving" has a more mathematical flavor because we talk about solving a set of equations, like the Newtonian equations that govern the motion of an object. In both formal systems symbols are being manipulated according to a set of rules. Not so with Tadro. With the exception of what's going on in its microcontroller (see Figure 5.1), Tadro manipulates only its body as it interacts with the rule-based physical world.

Just because we can represent physical rules—by computing physical interactions within and among physical entities—doesn't mean that the world is hosting physical interactions in the same way. Borrowing a page from Searle now, we'd say that the computer is not having the actual physical interactions but is, instead, simulating them via symbol manipulation.

By saying that the body does the "computational work," what Trimmer means is that the ongoing body-environment interaction, by virtue of its being an actual physical phenomenon, doesn't necessarily need to be mediated through a nervous system. From the neurocentric perspective (see Figure 5.2), the brain doesn't need to control how the tail interacts with the water because brainless physics governs that interaction. The brain doesn't need to solve Newtonian equations of motion. The physics takes care of itself according to its own rules. Without a neural imperative to "control behavior," what, therefore, does a nervous system need to do?

ARE BRAINS COMPUTERS?

It's not that brains are unimportant. Brains do something—when they are present. The paradox is that some behaving animals and robots don't have any structure or program that we would say is a "brain." But before we talk about brainless behavior, we need to delve deeper into what we think brains are and what we think they do.

A huge body of physical evidence shows that the embodied-brain in a variety of animals is involved in some of the functional events that create the behavior that we recognize as an agent interacting with its environment. Are we happy now? Isn't this what we've been intuiting all along about the importance of brains? No, no. Academics are never happy because the world is never that simple. And what brains do is not simple.

Let's go back to the contrasting paradigms of Turing and Searle. Turing gets the blame (or credit) for this whole "brain is a computer" problem, having argued that if every kind of thought that a human might have was an algorithm—where an algorithm is a mathematically expressible series of instructions for completing a specified task—then a computer was working IN THE SAME WAY as a brain, manipulating symbols in a deterministic manner.[23]

In case you missed my subtle use of capitalization, the key phrase here is "in the same way." This gets us to the heart of the matter: if two different types of physical contraption are operating "in the same way," does that mean that they are the same thing? For example, if a coal-fired locomotive and a diesel automobile both operate by expanding gases in a chamber and using that pressure-volume work to push a cylinder, are they the same thing? On the level of pressure-volume work transduced to linear displacement, yes. On another, no. The locomotive heats a boiler filled with water that is turned into steam; the automobile compresses the vaporized diesel fuel, which then explodes. My ten-year-old daughter would also point out that locomotives run on tracks, whereas cars run on the road; locomotives pull huge numbers of cars behind them; automobiles are smaller and have rubber tires.

Several of you reading this are snickering, I can tell, because you love trains and have thought of something wicked to disturb our little thought experiment: turn the coal-fired locomotive into a diesel-powered one, just like the automobile. Now locomotive and automobile use power plants that operate in an identical fashion. Size still a problem? You may have ridden on small-scale trains that are about the size of automobiles (or buses). Tires and tracks? You see the game: in the face of objections to sameness, change the objecting feature to be the same, *ad infinitum*.

What we did initially was to focus on what we saw as an essential property of our system—the power plant—and let that drive our discussion of sameness. This is akin to the process of making analogies: finding the similarities between two situations or mental representations and then using the similarities as a reason for inferring something new about one of the two entities.[24] Because both locomotives and automobiles are the same in terms of using diesel-fired internal combustion, we infer, by analogy, that they must have other similarities. And they do, depending on how we construe the categories for those other similarities.

Because the mental categories that we create are hierarchical and, if we are careful, mutually exclusive at a given level in the hierarchy, we can always find similarities among any and all physical things.[25] This kind of general sameness—for example, all physical things are similar because they are composed of matter—is what a philosopher would call "trivial." Trivial inferences are either tautological (as here, when we define the thing of interest by merely restating the thing) or self-evident.

Lesson learned: whenever we talk about two things being the "same" or "different," we need to first (1) define our terms and agree on those definitions, (2) define and identify our categories of comparison, and then (3) keep the discussion focused on those terms and categories.

Let's try this out. By analogy and category expansion we can always argue that a brain is the same thing as a computer. However, can we find some nontrivial category of in-the-same-way-as, some functional equivalence between brains and computers? Yes, we can. Both brains and computers can use, at least some of the time, explicit algorithmic steps to calculate. By "calculate" here I mean, as we were talking about with a Turing-computable algorithm, steps of instructions that can be explained with symbols such as numbers or words. In addition, those instructions, if followed exactly, will always produce the same result if you start with the same inputs. Thus, in the category of "calculators," brains and computers can perform mathematical calculations like "1 + 1 = 2." With input from external sensors brains and computers can perform more complex calculations like "when will that football hit my head."

Now you might argue that calculation is a trivial, self-evident similarity because humans created digital calculators, what we call "computers," with their brains. Thus, you'd argue, we merely figured out how our brains worked and then made machines do the same thing. Precisely. But far from being trivial, I'd argue, this is an example of what philosophers of mind and artificial intelligence call "functionalism," the similarity of how two different entities operate.[26] Under functionalism, "the mind" of humans contains a whole library of innate and learned functions that can be carried out by any number of physical mechanisms, including a brain and a computer.[27] From this perspective, the brain is a computer and the computer is a brain in the sense that both can work the same way—manipulating symbols— at least some of the time and for at least some kinds of operations.

Emotional response: any damn fool knows that a brain and a computer are different! Just look at them. One is grown out of cells, is wet, and is possessed by animals. The other is built by human animals out of silicon, metal, and plastic and it sits on my desk. They may share some functions—like calculation and memory—but I don't see my computer functioning as a food-seeking, reproducing, and evolving creature. Hmmm . . . or do I? Maybe my computer just needs a body the same way that my brain does.

BACK TO BRAIN BASICS

Confused? That's a good sign, even if you feel bad about the situation. Confusion means that your assumptions are being challenged and that you are open, probably against your will, to learning something new. But confusion is a vulnerable state. My students in "Introduction to Cognitive Science" hate confusion and the vulnerability it reveals. They came to college for information, not questions, damn it! How can they show me how smart they are if they don't have facts to learn, apply, and brandish? My job, they tell me, is to illuminate, not obfuscate.

Here's part of the problem: functionalism complicates the landscape, removing minds and intelligence from the sole province of humans. With the mind-blowing idea that intelligence might be found and built in nonhuman entities, most students seek comfort food. From their menu they order neuroscience: "Look," they say, "if we start from a neuroscientific perspective, we can build up from neurotransmitters, synapses, and neural circuits to an understanding of how brains, and human brains in particular, work. And once we know how human brains work," goes their logic, "we can understand what intelligence really is!"

Okay, folks. Let's see how far we get. Let's embrace the powerful neurocentric and reductionistic view (I'm not being sarcastic—it is powerful) and build a brain from the ground up. This is science, after all, and as we saw in Chapter 3 with the engineers' secret code, if we understand it, then we should be able to build it.

"But first," asks the evil professor, "what is the 'it' that you want to build?" Is "it" the brain? Which brain? If you mean an "animal brain," then, which group of animals? If you mean "human," then I have to ask, "What anatomical structures are you including?" What if our understanding of the human brain, at any structural or functional level, is incomplete? (Which it is.) Do you want to start instead with the more inclusive "central nervous system," which includes the spinal cord and its anterior extensions? Would you want to include the so-called peripheral nervous system? What about the sensory systems, including the proprioceptive systems in our joints and muscles? And hormonal systems, including the glands that make chemicals that alter brain function? And what about the circulatory system, which delivers those chemicals, including glucose and oxygen, to the nervous system? Do we include the lungs, which provide the oxygen, and the digestive system, including the liver, that provides the glucose?

Have I left anything out? No, and that's a blessing and curse. By admitting that that brain, however we define it and whatever structural context we put it in, is dependent on nonbrain stuff, we've just described an embodied-brain, a brain integrated into the whole critter. Can we dissect out just the brain? Yes. But as Phaedrus pointed out in *Zen and the Art of Motorcycle Maintenance*,[28] when you cut any system apart with your intellectual scalpel—when you analyze—you do so arbitrarily. In our case, when we look for nice, clean, predefined borders between "brain" and "body," there simply are none. The best we can do is acknowledge what analysis does and make it clear to ourselves and to others that there are an infinite number of ways to skin Weiner's cat.[29]

One place to apply the analytical knife is between anatomy and physiology. From the anatomical (structural) perspective, the brain is what it's made of. If we start in the most general way possible, we need an anatomical definition of "brain" that works for all animals. Richard and Gary Brusca, expert invertebrate zoologists, state, "[The] central nervous system is made up of an anteriorly located neuronal mass (ganglion) from which arise one or more longitudinal nerve cords."[30] This anterior (= toward the front of the animal) mass of neurons is called either a "brain" or an "anterior ganglion," depending on whom you are speaking to.

The brain of vertebrates, suggests Georg Striedter, associate professor of neurobiology and behavior at the University of California at Irvine, can be defined in three distinct ways anatomically: by region, by cells, and by molecules.[31] By region, Striedter says, "All adult vertebrate brains are divisible into telencephalon, diencephalon, mesencephalon, and rhombencephalon." By cells, the brains of jawed vertebrates possess the cells of the broad types known as neurons and glia. By molecules, Streidter explains that the brains of both vertebrates and invertebrates are characterized by the presence of the same neurotransmitters (you'll hear more about what those do later), including glutamate, GABA, acetylcholine, dopamine, noradrenaline, and serotonin. For the anthropocentric, Striedter points out that the anatomy of the human brain is different from other vertebrate brains by degree: it is more than four or five times larger than expected for a mammal of its body mass (a metric called "relative brain size") and has the most layers of neurons in the cortex (a division of the telencephalon).

If you think the anatomical perspective is messy, then put on your wet-weather gear as we approach physiology. From the physiological (= functional) perspective, the brain is what it does. The problem is that the brain, this anterior ganglion, "does" or participates in nearly every function of a vertebrate. Muscle contraction? Absolutely! Heart rate? Yes. Growth and development? Yes, even that, given the brain's involvement in hormonal regulation.

With all this brain-mediated physiology going on, it's extremely useful to try to focus on a single function. Let's go back to intelligence. Problem: we can't even agree on a definition. We've skimmed through the Turing-versus-Searle debate—that is just one axis of the pool of arguments. Even if we stick with skills-based definitions, we fight over what abilities indicate intelligence. Howard Gardner, professor of cognition and education at Harvard's Graduate School of Education, famously framed "multiple intelligences" for humans and, even though he has no fixed definition of intelligence, identified eight domain-specific types: spatial, linguistic, logical-mathematical, bodily-kinesthetic, musical, interpersonal, intrapersonal, and naturalistic.[32] Of course, Gardner's approach gets hammered too. So the paradox is this: how can we study a "thing" if we are all studying different things?

THE NEURAL CIRCUIT AS THE FUNDAMENTAL SENSORY-MOTOR SYSTEM

This is where neuroscience really shows its powerful Kung Fu. See, grumpy students of mine, I actually agree with you that by starting with the basics of nervous systems, we create an empirical and materialistic foundation, a molecule → cell → region → body chain of causal understanding.[33] However, this understanding only takes us so far, at the moment, and then we are left holding the bag of subjective first-person and other-minds experience to intuit an understanding of our human intelligence. The empirical promise, articulated cogently by

Patricia and Paul Churchland, professors of philosophy at the University of California at San Diego, is that our burgeoning neuroscientific understanding is creating a new and neuroscientifically based psychology.[34] Another important build-it-up-from-neurofundamentals approach is that of Jeff Hawkins, founder of Palm Computing and Handspring, who has demonstrated that the anatomy of the human cortex reflects the physiology of two fundamental pieces of intelligence: memory and the ability to predict.[35]

By carefully reviewing cortical anatomy and physiology, Hawkins has shown that the two are conjoined and inseparable. As Stephen Wainwright and Steven Vogel, cofounders of the field that we now recognize as comparative biomechanics, wrote in one of their early lab manuals, "Structure without function is a corpse; function sans structure is a ghost."[36]

Speaking of ghosts, one kind of confusion about the difference between a brain and a body stems from the implicit substance dualism that permeates our human cultures. Substance dualism, formulated carefully by René Descartes, claims that the mind is made of stuff that is different from the physical stuff that makes up bodies. Consider that for many people, a "mind" is equivalent to or created by a "brain." If, by intuition or religion, you believe that minds are made of a mysterious and nonmaterial substance, existing in some other dimension or plane after we die, then our brains must be of some special nonbody substance or quality too or possess the ability, as Descartes suggested, to interact with the nonphysical realm. Meanwhile, bodies, fashioned from clay, are earthly containers that are secular, ephemeral. The reasoning and predictions of substance dualism, however, have been refuted repeatedly.[37] We proceed knowing that brains and bodies are both physical entities, but we appreciate that the ghost of dualism lingers.

In practice, to study the brain scientifically we have to make some choices. If, for the moment, we choose to limit ourselves to just looking at sensory-motor neural circuits, what matters is the anatomical pattern of connections between neurons and the physiology of the type and timing of the chemical and electrical signals operating within the circuit.[38] To understand how any circuit functions, you also need to be able to measure when the circuit is active relative to when the animal possessing that circuit is doing something. Once you establish this correlation between a circuit's activity and the animal's behavior, then you need to test whether or not that circuit is necessary and sufficient for that behavior. You can test for necessity by removing the circuit genetically or surgically and then seeing how behavior changes.

Sufficiency is much harder to show: the activity of the circuit, independent of other circuits, must be able to cause the behavior previously correlated with the circuit's activity. To achieve the isolation that sufficiency demands, often the only way to go is to simulate the circuit on a computer or in an autonomous robot. The problem with simulation, as we've seen throughout this book, is that critics see simulation as "only" simulation, a model and not the thing itself.

Another way to show that a neural circuit is sufficient for a specific behavior is to find a "simple" animal—usually an underappreciated and overworked invertebrate sea slug, nematode worm, or fruit fly—that has the behavior and the circuit of interest but doesn't have all the other neural machinery vertebrates possess to complicate the analytical situation. The great power of the basic brains of invertebrates is that we can identify each neuron and its connections, something that remains nearly impossible, in practice, in vertebrates. With just a few overlapping circuits operating to move, find food, and mate, invertebrates have become powerful tools for neuroscientists. Using invertebrates as model organisms, neuroscientists have identified multiple circuits that are necessary and sufficient for escaping, digesting, flying, and learning.

Neural circuits get linked to behavior in two related fields called "behavioral neuroscience" and "behavioral neurobiology." Thomas Carew, professor of neurobiology and behavior at the University of California at Irvine, has made the strong argument that invertebrates help us understand the basic principles of anatomy and physiology that create the neural circuits that are necessary and sufficient to explain how animals behave.[39] And using general principles gleaned from invertebrates, along with the experimental approaches outlined above, neuroscientists are able to understand some behaviors in vertebrates. The most thoroughly understood behaviors, with mechanisms examined at the molecular through the behavioral levels, are echolocation in bats, hunting in owls, and navigation in rats.[40]

What behavioral neuroscience shows beautifully is just what Trimmer had said: every brain has a body. Once more, with feeling: understanding behavior involves not just the neural circuit but also the neural circuit placed within the nervous system, the nervous system connected to sensors and muscles, the sensors and muscles part of a particular body, and the particular body interacting with the physical world, including other agents.

We haven't, I realize, built a brain from the ground up. But by starting with circuits, we are making progress in terms of understanding how brains operate. By combining the brain basics of behavioral neuroscience with the functionalism of artificial intelligence, we come to three inescapable conclusions:

1. Every brain has a body, both in terms of cooperational physiology and connective anatomy. The brain alone is not sufficient to explain behavior.
2. The embodied-brain has some functions that it shares with computers and microcontrollers and some that it does not.
3. Some kinds of functions that we associate with the structure called the vertebrate brain we can see in so-called simple[41] organic and artificial agents that have no brain; thus, the brain doesn't control or determine all behaviors. The brain is not necessary for behavior.

This last assertion is probably the most controversial. Begging to differ might be George Lakoff, who helped develop the concept of the embodied mind within the fields of philosophy and cognitive linguistics.[42] Lakoff, writing about his development of the Neural Theory of

Language, states, "Every action our body performs is controlled by our brains, and every input from the external world is made sense of by our brains. We think with our brains. There is no other choice."[43] As you can see, Lakoff's embodied perspective is still a neurocentric one (see Figure 5.2). So let's get rid of the brain altogether and see what happens!

EMBODIED INTELLIGENCE: WHO NEEDS A BRAIN WHEN YOU HAVE A SMART BODY?

If, as Trimmer says, the body interacting with the world is doing part of the computational work of the nervous system, then we ought to be able to see bodies with very little brain or even no brain doing interesting things as autonomous agents.[44] Sound familiar? Doing just that is Tadro3, as I claimed at the beginning of this chapter. So let's continue to use Tadro3 to see how far we can push the idea of being intelligent without having a brain. I'll try to convince you that Tadro3 approaches the limit of being the simplest autonomous agent possible. By pushing the limits, I hope to show you that all it takes is a little KISS to create intelligent behavior.

You've seen Tadro3's neural programming (see Figure 5.1). Let me translate the central computation from its computer code into mathematical terms so you can see how simple its programming is. The computer code takes a voltage input from Tadro3's single eyespot and converts it to an intensity value, i, that is, in turn, converted to a value, β (Greek letter beta), which represents a turning signal for the tail, the tail angle:

$$\beta(t) = i(t) \times c$$

where the t indicates that both i and β are changing through time, t, and c is a "constant of proportionality," a numeric fudge factor that scales the light intensity to the size that we need to calculate a realistic tail angle. In words, this equation can be read as follows: "The angle of the tail at any time is linearly proportional to the intensity of light hitting the photoresistor at any time." That's it. It's hard to imagine a much simpler equation with variables. (I know, if you got rid of the c then it'd be even simpler, but at that point you'd have a simple identity equation.)

As brains go, this doesn't qualify. If we built a circuit that would perform this computation, you'd likely see something like this in a vertebrate (Figure 5.3): a sensory cell with a membrane potential that varies continuously with light intensity, a primary sensory neuron that converts the sensory cell's graded input into a train of action potentials and connects to two other neurons, one an inhibitory interneuron that reduces the activity of the motor neuron connected to the left-turning muscle and the other a motor neuron, without an intervening interneuron, connected to the right-turning muscle. This simple circuit has only four neurons and three other cells that complete the sensory-motor system.

Let's make the circuit even simpler! We can think about how a bioengineer might try to accomplish the task using wetware, cells and proteins of biological origin that she can arrange as needed. In her build-a-brain workshop she could make the circuit simpler by creating a receptor that connects directly to the muscle cells without any neurons at all (Figure 5.3). Let's presume that this simple circuit is, in principle, possible. Then this bioengineering design raises an awkward question: why have vertebrates made such a muddle of their circuit design? Why don't they go all the way with the KISS principle?

To put it another way: why go through the trouble of building a chain of multiple cells? There are good reasons. What you get with more neurons is more synapses. Each synapse, because it converts electrical signals to chemical ones, is a place where you can regulate and adjust how a neuron or muscle is responding to the "upstream" cell signaling the "downstream" cell. These cell-level adjustments are important in creating functional changes of the circuit during development and learning. Another consequence of having multiple neurons is that you can increase the number of connections that the

How a vertebrate might build
Tadro3's nervous system —

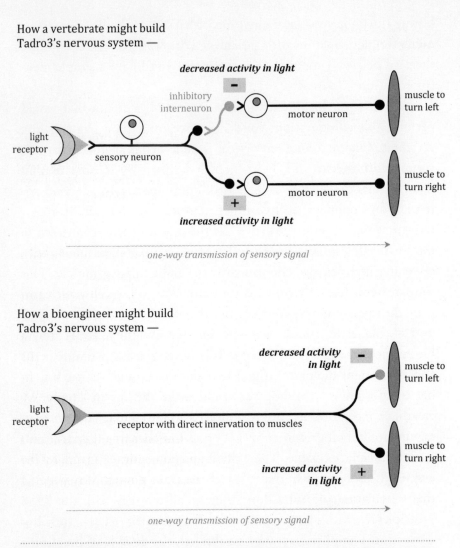

one-way transmission of sensory signal

How a bioengineer might build
Tadro3's nervous system —

one-way transmission of sensory signal

FIGURE 5.3. *Designing the nervous system of Tadro3 in wetware.* The top circuit, built in the way that vertebrates build neural circuits, contains seven cells: one receptor, one sensory neuron, one inhibitor interneuron, two motor neurons, and two muscle cells. The bottom circuit, built in a way that a bioengineer might be tempted to try, contains only three cells: one that directly innervates the two muscle cells. Both hypothetical neural circuits have the same function: in the presense of light on the receptor, decrease the activity of the left-turning muscle and increase the activity of the right-turning muscle. The gaps between the cells represent synapses, across which cells communicate by diffusing chemical neurotransmitters. A synapse is excitatory if unlabeled or labeled with a positive sign. A synapse is inhibitory if labeled with a negative sign. The large circles with smaller embedded dark circles represent the cell bodies of the neurons.

Tadro3 as a wheeled vehicle

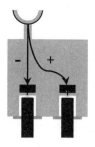

FIGURE 5.4. *Tadro3 morphed into a wheeled vehicle.* The single light sensor (cup) sends a reverse (−sign) and a forward (+ sign) signal to the two motors (small black rectangles) that independently control the two wheels (large black rectangles) that spin at different rates.

circuit makes with other circuits (branching connections not shown), increasing opportunities for coordination and computation.[45]

If you look at the neural circuits in Figure 5.3, you'll notice that we've done it again: we forgot the body! To be fair, we did this on purpose so that we could see what an isolated Tadro3 nervous system might look like. Note, also, that this circuit is not a brain in the anatomical sense of Brusca and Brusca's invertebrates: it is not a mass of neurons. This is a diffuse nervous system, and it needs a body. We can create a body that is as similarly abstract as the bioengineer's neural circuit. To keep it simple, let's put the Tadro3 nervous system into a wheeled vehicle (Figure 5.4). With a body operating on land, by the way, we don't have to worry about all the crazy physics of swimming that we mentioned previously.

Let's give Tadro3 simple wheels. The simplest wheels spin but don't turn. Tadro3 turns by having different levels of power go to the two motors that drive the wheels. Having two motors to turn in the wheeled Tadro3 is the functional equivalent of having two muscles, working in pairs, to control the direction of the tail for turning in the swimming Tadro3 (Figure 5.3). Also notice that the simplest neural circuit is used here: a single light sensor provides both an excitatory and inhibitory signal. The cup-shaped light sensor is directional in the sense that it registers light only when that light hits its concave surface directly (not by coming through the back of the cup).

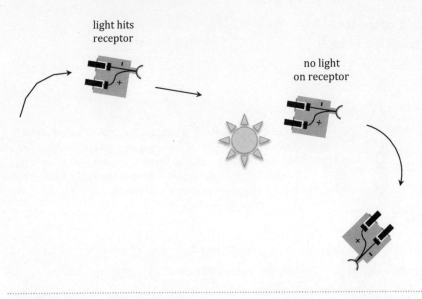

FIGURE 5.5. *Vehicular Tadro3 (vT3) turns in response to light.* In this thought experiment, when no light is hitting the cup-shaped light receptor, vT3 arcs to the right. When the receptor faces the light and is close enough to register the light, its path straightens as more power is delivered to the right wheel's motor and less is delivered to the left wheel's motor. The vT3 is inspired by the vehicles of Valentino Braitenberg.

How does vehicular Tadro3, or "vT3" for short, behave? It's time for a thought experiment, a cognitive simulation (Figure 5.5). First, suppose that vT3 has an intrinsic rate of wheel spinning and is always moving around. When vT3 is in the dark, the left motor gets a bit more power than the right, so vT3 arcs to the right. As soon as light falls on vT3's light sensor, though, the right motor starts to get more power, and the left gets some of its power reduced because of inhibition. When this happens, vT3 straightens out its heading.

Doing thought experiments like this, with a wheeled vehicle and a simple sensory-motor circuit, was the brainchild of Valentino Braitenberg, a neuroanatomist. His 1984 book, *Vehicles: Experiments in Synthetic Psychology*, inspired a generation of workers in artificial intelligence and behavior-based robotics. By taking the reader through thought experiments with an evolving fleet of Vehicles, Braitenberg creates the "law of uphill analysis and downhill invention." This law is drawn from what we would be tempted to call a functionalist observation: "It is actually impossible in theory to determine exactly what the hidden mechanism is without opening the box, since there are always many different mechanisms with identical behavior." [46]

Braitenberg calls analysis "uphill" because "when we analyze a mechanism, we tend to overestimate its complexity."[47] From Braitenberg we can see that anyone interested in understanding the mechanistic basis of behavior, including the behavior that we call intelligence, either has to open the box, as we did with our Searle hat on, or, as Braitenberg did, take the "downhill invention" route and create behavior from the ground up, like an engineer applying the secret code.

When we morphed Tadro3 into vT3, we showed that at least two mechanisms, two neural circuits in our case, could drive the sensory-motor responses to light. We also used Braitenberg's approach to show how little—in terms of circuitry—that Tadro3 needs in order to behave. Nowhere in the circuits of Tadro3 or vT3 do we see a collection and connection of interneurons that we'd be tempted to call a brain.

Braitenberg Vehicles are brainless and yet still manage to exhibit what, to the observer of the Vehicle who is blind to the Vehicle's internal mechanism, we would call intelligence, at least at the level of goal-directed, purposeful autonomy. To be fair, though, we haven't shown that vT3 actually works; we only ran the simulation in our minds. For part of his senior thesis in cognitive science at Vassar, Adam Lammert implemented the vT3 circuit on a wheeled robot to see if it worked as we had imagined. It did (Figure 5.6).

Because embodied Braitenberg Vehicles work both in physical reality and in simulation, we can use them to explore what kinds of bodies make behavior and what kinds don't (Figure 5.7).

In this embodied view you can see right away what's needed to be an autonomous agent. With a single sensor and a single wheel, what Braitenberg called a Vehicle of brand 1, this simplest autonomous Vehicle will speed up if it is facing a light source and slow down if not.

vT3, embodied and operating

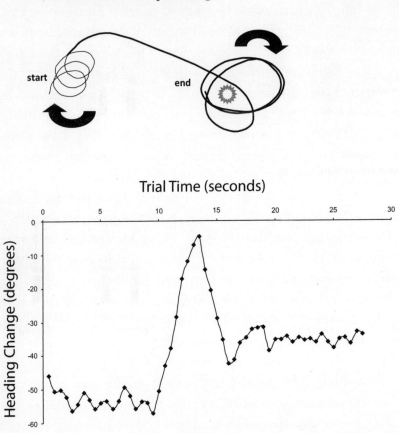

FIGURE 5.6. *The vT3 operating as an embodied and autonomous wheeled robot.* Top panel shows the arc-like path of vT3 over the course of a three-minute experiment run by Adam Lammert. When the path turns from gray to black, vT3 has detected the light. The bottom panel shows what happens when vT3 detects the light. At about ten seconds into the trial vT3 detects the light and changes its heading by almost 55 degrees, straightening out its arc, heading toward the light, and then orbiting it. Keep in mind that vT3 can be thought of as a Braitenberg Vehicle of brand 1.5.

three ways to create non-autonomous vehicles
that lack behavior

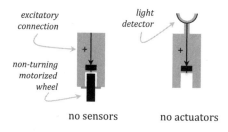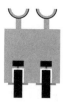

excitatory connection

light detector

non-turning motorized wheel

no sensors   no actuators   no connections

five ways to create autonomous vehicles
that possess different behaviors

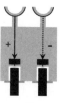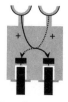

| link sensors to actuators | add sensors and/or actuators | change the type of connection | alter the pattern of connections | increase the number of connections |

parameter space for vehicles

1. sensors	0, 1, or 2
2. actuators	0, 1 or 2
3. connections from a sensor	0, 1 or 2
4. connections to a motor	0, 1 or 2
5. connection type	excitatory (+) or inhibitory (-)

FIGURE 5.7. *The embodied view: intelligence is what we do autonomously.* In a thought experiment created by Valentino Braitenberg, simple vehicles can have sensors attached directly to actuators, without an intervening brain. Without a sensor, an actuator, and a connection between them, the vehicle cannot behave because it has no way to sense or move. To have autonomous behavior, a sensor must be connected to an actuator. Here is a simple thought experiment: take the autonomous vehicle with one light sensor, one motorized wheel, and an excitatory connection between them. Put a light in front of this vehicle. What happens?

Although this isn't terribly exciting behavior, it is behavior as we've defined it: the interaction of the agent and the environment. Vehicle 1 shows that what's necessary and sufficient for behavior is (1) a sensor connected to a motor, (2) the sensory-motor linkage embodied in a chassis that has an actuator, (3) the Vehicle situated in an environment with a variable energy field that the sensor can detect, and (4) the Vehicle situated in an environment with a substrate to which the actuator can transfer its momentum. Behavior is impossible if any of these features are missing.

If you study these Braitenberg Vehicles (Figure 5.7), you can see where vT3 might belong: between the first and second autonomous Vehicles. In the lexicon of Braitenberg vT3 is thus neither a Vehicle of brand 1 (single sensor, single motor) nor a Vehicle of brands 2 or 3 (double sensor, double motor). In recognition of its intermediate character, Lammert called vT3 Vehicle 1.5. We can characterize Vehicle 1.5 not only graphically (Figure 5.4) but also by using the parameter space for Vehicles (Figure 5.7): (1) one sensor, (2) two actuators, (3) two connections from the one sensor, (4) one connection to each motor, and (5) both excitatory and inhibitory connections.

EMBODIED AND SITUATED AGENTS

No brain? No problem. As Tadro3, vT3, and Braitenberg Vehicle of brand 1.5 all show, we can build autonomous agents without what Professor Rodney Brooks of the Massachusetts Institute of Technology calls the "cognition box." Brooks, a mainstream member of the world of artificial intelligence, revolutionized AI in the 1980s. While others had slow-moving robots burdened with computationally intensive problems like vision, path planning, and world mapping, Brooks built simple robots that could literally run circles around their more complex brethren.[48]

Inspired by what invertebrates could do without much in the way of a brain, Brooks and his colleagues programmed the computers inside mobile robots with a parallel of arrays of what most of us would call reflexes. In a reflex, a simple stimulus, like intense heat on the palm of your hand, causes an immediate response: flex the joints of your arm. When the joints flex, your hand moves toward your body and usually away from the heat source. In this sense Tadro3 also works by a kind of reflex, one that is ongoing and gradual rather than working from an on-off switch.

Brooks reasoned—and then demonstrated in the mid-1980s—that robots could use a storehouse of reflexes to do what the brain-based, cognition-box robots of the day could not: navigate in a changing environment. Brooks's autonomous six-legged robot, Genghis, could walk over rough terrain and follow a human.[49] At the time Genghis was a breakthrough in the true sense of the word, the existence proof for what has become the field of behavior-based robotics.[50]

Behavior-based robotics uses the synthetic method to build up from the basics. We've encountered this when we spoke of Braitenberg's "downhill invention" approach to understanding behavior. The synthetic approach also works hand in hand with the KISS principle because the whole idea is that the building blocks, like the reflex modules, are simple constructs, as with a stimulus linked directly to a response. The synthetic approach also works with our secret engineers' code because we can understand the simple elements and then, piece by piece, put them together to build other now-more-complex systems that we can still understand.

When you start to synthesize the nervous system of an autonomous agent out of reflex modules, you run into an immediate problem: how do you coordinate those modules? If each reflex module automatically creates a behavior when its stimulus switch is flipped on, then what happens if two behavior modules get flipped on at the same time? Or what happens if behaviors are stimulated in sequence, one after the other, and their automatic actions overlap in time? This kind of conflict between automatic controls needs to be settled by an arbiter, a system that decides which module gets the green light. Brooks, again inspired by animals, created an arbitration scheme that he called "subsumption." In a subsumption-style neural architecture, the robot's programmer ranks the behavior modules. In case of conflict, the behavior module with the higher rank "subsumes," or suppresses, the behavior module with the lower rank.[51] Once programmed, subsumption is a built-in decision arbiter. You, the autonomous agent, don't need to consider what to do next; you simply do the lowest-level behavior as the default until you are stimulated to do something else. I've tried to program myself to operate with subsumption when I drive. At the bottom of the hierarchy, my default layer is a behavior I call "drive efficiently." This module is actually a collection of submodules that include behaviors like: avoid sudden acceleration, adjust speed to avoid red lights, and choose uncongested routes. At the top of my two-level subsumption hierarchy is "drive safely." This module is a coordinated set of modules straight out of driver training class: stay on the road, keep a safe following distance, don't hit the car in front of me, and scan ahead for possible problems. In practice, the "drive safely" behavior overrides "drive efficiently" most of the time because the presence of other cars or challenging driving conditions like rain, darkness, or unfamiliar roads stimulates the module.

With subsumption in mind, I enjoy trying to analyze the driving behavior of other humans. At the lowest level many drivers appear to have the behavior, "drive like hell." This default appears to involve a collection of submodules that includes behaviors like: pass or tailgate any car in front of you; switch lanes rapidly and, if necessary, without signaling; prevent other cars from passing you; accelerate quickly from a stop. At high-stimulus thresholds, in most drivers the "drive safely" module appears to override "drive like hell."

Any agent who can do the behavior arbitration dance with subsumption must have some familiar accoutrements: a body, a body with sensors, a body with actuators, and a body operating in the real world. Brooks sums up these requirements as follows: an autonomous agent must be embodied and situated. An embodied agent reacts to events in the world by virtue of having a physical body; this is the "body computation" that we've talked about before. A situated agent reacts to events in the real world by virtue of having senses; this is the basis of the "neural computation" that we invoke the minute we put together a circuit diagram of a nervous system.

TADRO4, FINALLY: TO EAT AND NOT BE EATEN

As you've probably figured out, Tadro3 lacks a subsumption-style nervous system. Tadro3's decisions are all ongoing, continuous adjustments of the turning angle of the tail. The resulting light-seeking behavior gives Tadro3 the know-how to detect light, move toward it, and then orbit around the spot of highest intensity.

Sadly, even though Tadro3 can evolve better feeding behavior by evolving its body, it lacks the genetic wherewithal to evolve different skills. For example, if danger is lurking, Tadro3 has no way of knowing: it just senses the intensity of light through its single eyespot. See no evil, hear no evil. Such no-know-how is a good way for an organic agent, like the tunicate tadpole larva after which Tadro3 is modeled, to become lunch in the game of life.

Predation is thought to be one of the strongest selection pressures in living fishes, as we talked about at the end of Chapter 4. Applying a strong and ecologically relevant selection pressure like predation thus seems like a great way to get back to where we started: trying to understand what drove the evolution of vertebrae in the first fish-like vertebrates.

If predation is the hypothesized selection pressure, then, for the reasons just mentioned, Tadro3 can't do the job. It's not built to be prey. Instead, we need to upgrade to a Tadro that has both the nervous system and the body to eat and to avoid being eaten. Tadro4 is up to the task, and I'll explain its design using the ideas we've developed in this chapter on embodied intelligence.

We designed Tadro4 to do what living fish (but not tunicate tadpole larvae) do. Tadro4 swims around with two eyes (photoresistors) foraging for food. When and if a predator approaches, Tadro4 detects the predator using an infrared proximity detector, which is the functional equivalent of a lateral line—an array of tiny hairs and cells running along the length of the fish's body that move when water is displaced by the fish itself or by something nearby moving.[52] When a detector on either side of the body is triggered, Tadro4 tries to escape. This switch in behavior, from feeding to fleeing, is accomplished by a nervous system that is a two-layer subsumption hierarchy (Figure 5.8).

What's really cool about this two-layer subsumption design is that it very closely resembles, at a functional level (think: functionalism), how the nervous system of fish actually operates. Most fish find food by foraging—swimming around and searching for chow. This is layer 1, the default behavior, the kind of behavior that Tadro3 performed when we pretended that the light was a food source. In addition fish also are able to detect predators, and if a predator strikes, the would-be prey hits the neural panic button and performs what we creative biologists call a "fast start." This is layer 2, the behavior ranked higher in terms of importance that layer 1.

The fast start is an escape response that involves the highest accelerations ever measured in fish, over ten times the acceleration due to gravity.[53] For comparison, astronauts on the US space shuttle experience maximum accelerations of about three Gs when the main engines ignite for the last minute of orbit-reaching propulsion.[54] The simultaneous firing of nearly all of the muscles on one side of the fish's body make these incredible accelerations possible. This muscle activity is coordinated by a purpose-built neural circuit called the recticulospinal system.[55] The reticulospinal circuit activates the motor neurons of the muscle after it receives a stimulus from the eighth cranial nerve, the nerve that is connected to the inner ear and the lateral line of the fish. Sounds like predator detection if you ask me. Given that the lateral line runs all the way to the tail in many fishes, this is like having the proverbial eyes in the back of your head—or, um, body.

Here's the really cool part: if the fish is swimming around when it detects a predator, this escape-response neural circuit overrides the swimming-around circuit! That's subsumption, baby. This override was demonstrated in a series of elegant experiments on goldfish by Joe

Tadro4 switches behaviors

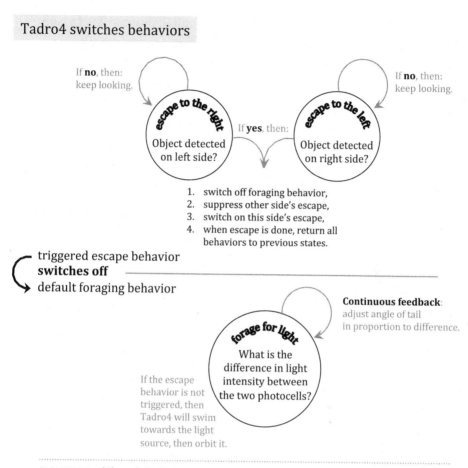

If **no**, then: keep looking.

escape to the right

Object detected on left side?

If **yes**, then:

escape to the left

Object detected on right side?

If **no**, then: keep looking.

1. switch off foraging behavior,
2. suppress other side's escape,
3. switch on this side's escape,
4. when escape is done, return all behaviors to previous states.

triggered escape behavior
switches off
default foraging behavior

forage for light

What is the difference in light intensity between the two photocells?

Continuous feedback: adjust angle of tail in proportion to difference.

If the escape behavior is not triggered, then Tadro4 will swim towards the light source, then orbit it.

FIGURE 5.8. *Like a fish, Tadro4 is built to decide when to forage for food and when to escape from predators.* The decision to switch behaviors is made using a two-layer subsumption architecture: Tadro4 forages for food until the escape response has been triggered, at which time the foraging is switched off until the escape is complete. Every sensor on Tadro4 can be thought of as continuously answering a question: where's the food (eyes)? Where's the predator (lateral line)? The specific answers provide the continually updated perceptions that alter the state of the embodied-brain and drive the immediate actions of Tadro4.

Fetcho, professor of neurobiology at Cornell University.[56] He and Karel Svoboda directly measured the nerves' activity in the fast-start and steady-swimming circuits. The neural signals for steady swimming comes from a system of so-called central pattern generators, clusters of neurons that drum along at a steady rhythm without much input from other circuits. Inputs from the fast-start circuit, though, immediately switch off steady swimming when escape is activated.

Our design of Tadro4 is propelled, if you will, by what we know about how living fish respond to predators in terms of the neural circuitry, swimming behavior, and evolution. Because we know so much on so many different levels, the predator-prey system is an excellent one for testing hypotheses about the evolutionary origins of vertebrae.

Vertebrae? Remember them? We've lost sight of these axial structures as we probed embodied brains and intelligent behavior. For Tadro4, we created an axial skeleton that had actual vertebrae. So we resolved to see how a population of Tadro4 prey responded, in terms of the number of vertebrae, to selection imposed by a predator. The game of life continues.

chapter 6

PREDATOR, PREY, AND VERTEBRAE

W E ALL KNOW WHY PREDATION IS SUCH A STRONG SELEC-
tion force: dying sucks. Those about to die, here to salute
you, are the Tadro4 prey, introduced in the last chapter as the feed-or-
flee Evolvabots. Even though the Tadro4s bear the name Tadro, they
are, compared to Tadro3, a different kettle of fish. The most important changes have to do with what Tadro4 and its whole new world
seek to model: vertebrates in a prey-versus-predator world. Tadro4 has
to be equipped with a new nervous system and a new body in order to
eat and try not to be eaten. And, to pull this all off, we also needed to
design and build a predator capable of tracking and chasing the
Tadro4 prey.

When you are in the middle of a big, multiyear project like the
Tadro3-athon, you have plenty of time to second-guess your no-
turning-back decisions. We'd regularly convene our whole team of Vassar
and Lafayette College researchers, give progress reports, teach one another about our different fields, and then try to find gentle ways to tell
our colleagues and students that their work was a piece of sh . . . , sh . . . ,

shaving cream. That's the charm and the curse of the academic: you are trained to criticize anything that moves. We are trained to be intellectual predators. As professorial types, we model this predatory behavior for our students, stalking not people but the ideas that they use and generate. We also model being the prey, the one presenting an argument to the circling sharks.

All students are prey. Sorry. When we are in student mode, we have to admit first to ourselves and then to others that we don't know something. Talk about vulnerability! Who wants to be an openly vulnerable prey item? But ignorance or misunderstanding is a problem: no way around it. So anytime we want to get better at what we are doing, we have to admit that we have problems. As Jenny Ming, cofounder and past president of the clothing retailer Old Navy, said, "You can't fix something if you don't admit it's wrong."[1]

PROBLEMS WITH TADRO3 HELP CREATE TADRO4

Okay. What's wrong with Tadro3? I've already taken you through our acute crisis of confidence in Chapter 4. When the population of Tadro3s wasn't evolving as we predicted, we switched our collective behavioral mode from conduct-experiments-and-analyze-data to stop-and-look-for-mistakes. We found the mistake that turned out to be a happy one: the wobble of Tadro3 was not a sign of energy inefficiency but rather a metric of enhanced maneuverability. That problem was tactical. On the strategic level chronic problems can't be solved in the same way: they are a by-product of making plans. Once a particular set of strategic plans are set in motion, chronic problems persist unless you revise and start over or until you evaluate the strategically flawed experiment and design the next one.

On the strategic level, you can argue, as my colleague in biology did way back in Chapter 1, that modeling itself is a conceptually flawed process. You won't be surprised, from everything I've said about modeling, that I disagree with the no-modeling critique.

At the same time, I agree with Barbara Webb (see Chapter 3): we need to justify carefully the specific modeling approach that we take.

So now that you've been through our design process (Chapter 3), seen the Tadro3 model system in action (Chapter 4), and gotten a taste of the mechanisms that help us understand Tadro3's behavior (Chapter 5), we can revisit Webb's criteria for judging the value of our biorobotic model.

By Webb's criteria, have we produced a good model? Recall that in addition to "relevance" (= able to test a hypothesis) and "medium" (= physical basis of robot), we selected "behavioral match" and "mechanistic accuracy" as two other figures of merit for our Evolvabot models. Do we see matched behaviors and accurate mechanisms within the hierarchy of skeleton → individuals → population? In terms of behavioral match, the complexity of the evolutionary patterns in the Tadro3 system is evidence that the system behaves realistically on the population level. Further, because we know that evolutionary mechanisms of selection and randomness cause those complex patterns, the accuracy of those mechanisms is also high.

But what about the mechanistic accuracy of the Tadro3 itself, the levels in our hierarchy of individual and skeleton? When we present our evolutionary biorobotic work at the Annual Meeting of the Society for Integrative and Comparative Biology, I let our students know that they can expect a visit from a constructive predator from the laboratory of Robert Full, professor and director of the Poly-pedal Laboratory at the University of California. Full is world famous for his careful experimental study of invertebrates, isolation of functional principles that unify seemingly different behaviors, and implementation of those principles in the design and operation of biomimetic robots.[2] So when Full or one of his colleagues arrives to critique, we are, as the Ferengi say, all ears.[3]

One of the most powerful critiques from Full's lab is that our Tadros are too simple. Therefore, the argument goes, the Tadros are not accurate models of biological phenomena. Tadro3 fails as a model at the level of an individual agent. As with all useful critiques, it's true, at least in part. So as prey items, eager to learn, we first try to admit that we have a problem. Then we try to figure out how—or if—to fix it. One defensive fix is to attempt to do a better job of explaining the different ways that you can judge biorobots as scientific models. That's why I summarized Barbara Webb's approach to biorobotic modeling in the first place.[4]

Another fix is to go back to the beginning and revisit the KISS principle. We used KISS to justify creating a simple model first. Now we were being criticized for the simplicity of Tadro3. So what's next? What new elements do we add to the model, and why? Will those new features, which will certainly make our robots more complex, also be the right ones to make our models more accurate mechanistically and in terms of representing biological systems? To answer the what-next question is why I took the time in Chapter 5 to talk about embodied brains, neural circuits, and the subsumption approach to modeling the recticulospinal sensory-motor system of fish. We add biologically based complexity to the nervous system of Tadro3 to create Tadro4.

But what about testing a hypothesis? Is the Tadro3 system relevant? Yes. We set out to use our Evolvabots to test the hypothesis that selection for enhanced feeding behavior drove the evolution of vertebrae in early vertebrates. We substituted structural stiffness of the notochord for number of vertebrae in a vertebral column. Because the population of Tadro3 evolved reduced structural stiffness but better swimming behavior under selection, we disproved the hypothesis that selection for enhanced feeding behavior drove the evolution of vertebrae. However, we came to recognize that the hypothesis itself is probably overly simplistic, given that feeding behavior and structural stiffness of the notochord can become decoupled with a fitness function that rewards multiple sub-behaviors for feeding alone. This led us to add complexity to the selection environment of Tadro3, tossing in a predator to create the wild and crazy new world of Tadro4.

Another set of critiques of Tadro3 came from the predators within. Every time we got our multi-institutional research group together, Rob Root, professor of mathematics at Lafayette College and one of the team leaders on the evolutionary simulation project, was always itching to sink his teeth into the robotic Tadro3. One of the many great things about collaborating with Rob is that his bites are gentle and always meant to be constructive. It also helps that he is persistent when we want to deny that we've got any problems. As we slowly unrolled the methods and results from the Tadro3 experiments over many months, here's the blood in the water that got Rob's jaws clacking: (1) our fitness function was a composite of feeding-related behaviors rather than being, for example, the actual amount of food collected; (2) our fitness function was a sum of unscaled measures; (3) the composite feeding behavior did not include acceleration performance; (4) the population size was too small (remember in Chapter 2 that Rob had brought up the issue of the large effect that randomness can have with a small number of cases); (5) the number of generations was too few to show large evolutionary trends, and (6) vertebrae were missing in action.

I ran these recollections by Rob recently. He agreed that these had been his primary critiques. He had always found it fascinating, he added, that vertebrae evolved independently multiple times in vertebrates. That convergent evolution, he noted, appeared to be contingent upon the prior evolution of enhanced sensory systems, notably the paired nose, eyes, and ears that characterize vertebrates. This observation is part of the justification for the sophisticated nervous system and sensor systems that we talked about in the last chapter: Tadro4, to succeed as a potential prey in the prey-versus-predator world, needs to have the nervous and sensory-motor systems to give itself a fighting chance of survival.[5]

At the time of Rob's critiques, all we could say, given that we couldn't change our evolving Tadro3 system mid-stream, was, "Great points, Rob! Well taken. Next time." Fortunately, at the same time that the robotic Tadro3 system was busy evolving, Rob and another one of our collaborators, Chun Wai Liew, associate professor of computer science at Lafayette College, were taking on the Tadro3 system and the biological hypothesis it tests using a different technique: digital simulation.

GOING DIGITAL ON THE WAY TO TADRO4

I'd done my own predatory critique back in Chapter 1, bringing up digital simulation only to dismiss it. The dismissal part was not entirely fair, I know, and it raised some hackles on my friends who gave me feedback on the early draft of that chapter. My point, which I still argue is valid, was that embodied robots have the advantage over digital ones because the embodied 'bots can't violate the laws of physics. This is true. However, if you can build an accurate "physics engine"[6] and use it in your digital simulation, then you can avoid creating completely unrealistic models. You can even learn something too, as we'll see.

To model a world that gives your actions predictability, you need the rules that a physics engine provides. Gravity, momentum transfer, and projectile motion are all implicit rules in role-playing video games like Grand Theft Auto. Entertainment, argues Tom Ellman, associate professor of computer science at Vassar, is just the tip of the iceberg for physics-based animation. Science, education, and engineering are other areas in which realistic world models are important. Although building situation-specific animations is difficult and time consuming, Tom has developed software that automatically, based on inputs from an interactive human user, unfolds the physics-based world.[7] Aeronautical and automotive engineers use real-time three-dimensional animation of rigid bodies to help speed the development of new airplanes and cars. When you can build and test a vehicle on the computer, then you can bend, twist, kill, crush, and destroy it with impunity. And you can do so over and over to explore the impact of changes in the vehicle's design.

As part of their design process using a digitally simulated world, engineers also employ what they call "genetic algorithms" (GA). The GA approach is evolutionary, using randomness to create novel variants of a design. The performance of the variants is judged using a fitness function that seeks to maximize, usually, a single aspect of performance. The variants with the best fitness are selected to mutate and mate to create a next generation of novel designs for testing.

By searching for an engineering design that optimizes a single aspect of performance, such as fuel economy, the GA process is an example of a class of procedures called "hill climbing" routines. The hill, in this case, is the specific area in your design space that combines features in a way that gives you the best possible—the optimal—results.[8] Once your evolutionary simulation finds the top of a hill, you then stop and build the winning design as a physical entity. By the way, only if that physical embodiment of the digital simulation works as predicted can your digital simulation be said to have been "validated." In many ways, we were validating backward, having started with a physical system, Tadro3, and creating the digital simulation of it.

In need of a physics engine for the Tadro3 world, Rob, Chun Wai, and their students Megan Cummins and Greg Rodebaugh took on the task. They had to simulate the force-coupled interactions of the elastic and flexible tail of Tadro3 with the water motion that the tail and body creates. No matter how simple "go build a physics engine for Tadro3" sounds, in practice the problem is a bear, for reasons mentioned in the last chapter when we were talking about using the body rather than the nervous system to "solve" all the hard computational problems. Here we actually had to grapple with all the mathematical complexity. Rob and Greg focused on a creating a tractable approach to the physics. They began by modifying the "immersed boundary layer" method created by Charles Peskin, professor of mathematics at New York University's Courant Institute, to work for the particular situation of swimming fish.

Meanwhile, Chun Wai, an expert in evolutionary computing, wisely proposed that while Rob and Greg were building the better physics engine, he and Megan make progress on the digital evolution front by using a less accurate but computationally straightforward mathematical model, developed by the late Sir James Lighthill, called the elongated-body theory (EBT). With help from Rob, they used EBT to calculate the thrust that the digital Tadro3 would generate. Once digital Tadro3s could wiggle their tails, generate thrust, swim, and turn, Chun Wai and Megan created an evolutionary world, one in which the digital Tadro3s (let's call them digi-Tad3s) attempted to detect light, swim toward it, and then orbit it (Figure 6.1).

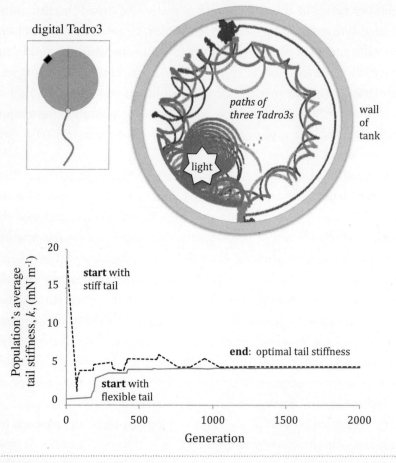

FIGURE 6.1. *Digital Tadro3s extend the experiments with robotic Tadro3s.* Although not embodied, Digi-Tad3s are self-propelled and autonomous, seeking light and then orbiting around it in a computer simulation. The paths of three Digi-Tad3s are overlaid (upper-right image) to show how they spin toward the light and then stay close to it. In the graph below, two populations of Digi-Tad3s start with very different average tail stiffness but then evolve, over about a thousand generations, to a similar stiffness. This convergence of the populations' average tail stiffness to the same value, no matter where they started, was consistent in over one hundred different populations. The existence of an optimal tail stiffness, a "hill" in the sense that we talk about in the text, leads us to make a prediction about the robotic Tadro3 world: the population of robotic Tadro3s may have started with a stiffness above the optimal; hence the evolutionary decline. The results from digi-Tad3 also cause us to realize that evolving vertebrae is not just about making the axial skeleton as stiff as possible. Stiffness has to be just right to meet the conflicting mechanical needs of propulsion and maneuverability.

Keep in mind this strange fact: digi-Tad3s are representations of a representation (tunicate tadpole → embodied Tadro3 → digital Tadro3).

Sound familiar? It should: the digi-Tad3 world modeled the physical world of the robotic Tadro3. Released from the time-and-person constraints of experiments in the real robotic world, we could test thousands of individual digi-Tad3s in each generation and, furthermore, test thousands of generations. Best of all, the digi-Tad3s gave us the ability to rerun the experiment hundreds of times, each time varying the tail stiffness with which the population started.[9] It's true: you just can't do that with embodied robots.

The most interesting result from all of the digi-Tad3 runs was this: no matter where a population of digi-Tad3s started in terms of average tail stiffness, the population evolved toward the same value of stiffness, what looks like an equilibrium (Figure 6.1, lower graph). We've called this equilibrium value the "optimal hill of tail stiffness" because it appears to balance the mechanical demand of rapid propulsion, on the one hand, and maneuverability, on the other. Although a very stiff vertebral column allows the digi-Tad3 to swim quickly, that same stiff-tailed digi-Tad3 can't orbit tightly around the light source. The balance between speed and maneuverability is enforced by the fitness function, which rewards increased speed and, at the same time, a decreased orbital radius.

What we've learned about the apparently optimal tail stiffness in Digi-Tad3 helps us understand good ol' embodied Tadro3. If you remember, two of the four times that selection was present in the population of Tadro3s, the tail stiffness increased. Tail stiffness of the population decreased the other two times that selection was present. We highlighted this pattern in Chapter 4 by pointing out this surprise: the same selection pressure generated two different directional trends. What Digi-Tad3 suggests about this oscillation is that the stiffness of Tadro3's tail was moving back and forth in order to keep climbing the hill and finding the equilibrium. In other words, we didn't see tail stiffness evolve in a single direction because we happened, by chance, to have started the population of Tadro3s next to the hill of optimal stiffness. Tadro3 was "born on the side of a hill."[10]

BACK TO THE EMBODIED

In both digital and embodied worlds the fitness function computes the selection pressure in each generation. The fitness function is the algorithm that we use to judge the players in the game of life. Differences in the performance of the players, coupled with the random influences of mating, mutation, and genetic drift, change the composition of the population of players generation by generation.

Given selection's central role in the evolutionary game, Rob's point about how we implemented the fitness function in Tadro3 was a great one. He suggested that we simply measure the energy that each Tadro3 harvested. Nick Livingston, working as the assistant director of the Interdisciplinary Robotics Research Laboratory at Vassar, hit upon the same idea. To measure the amount of energy harvested, Nick suggested that we mount a small solar panel on top of each Tadro and let the Evolvabots use the energy that they collect as a direct fitness function. Just let the Tadros go: "Fly! Be Free!"[11] The Tadro with the highest fitness would be the last Evolvabot standing, figuratively, or swimming, literally.

A solar-powered Tadro4 is a great idea for both strategic and tactical reasons. Strategically, we'd be using actual energy harvesting as a behavior linked directly to survival, just like organic agents do. Direct energy harvesting would eliminate the criticism that we were "just" simulating selection with the fitness function calculated from numbers assigned to a host of arbitrary behaviors. Tactically, direct energy harvesting would make our experiments run much faster. When we analyze video frame by frame and manually select the points of all our robots, we end up needing tens of person-hours of earnest effort to make it happen. Because we need the data from the video to calculate fitness and we need fitness to create the genomes of our next generation of Tadro3s, the video analysis is a bottleneck. Perhaps more importantly, we all go mad, mad, mad from the excruciating boredom of the manual process.[12] I'm always surprised that video analysis is not a stronger selection pressure on our students; they are really tough (some even claim, when pressed, to "not mind it"—these students are the Zen masters among us).

Direct energy harvesting for locomotion should work, no problem. Solar-powered cars race across the Australian Outback in the World Solar Challenge.[13] We took a page from their playbook and added a solar panel to a Tadro3. We wired the panel to provide electrical power to a battery. However, we ran, or swam, into an immediate problem: we didn't have enough solar power. Tadro3's water world, unlike the desert, has a limited supply of light, a concentrated source that is surrounded by near-dark conditions. We tried increasing the overall level of light, but then we quickly overwhelmed the light sensors, and they couldn't detect any differences. We added sunglasses (I'm not joking), but still no luck. Then we took a different tack. We gave the Tadros a small electrical charge to get going in the dark. However, in practice, each Tadro gets a slightly different amount of energy. This unfairness comes about because all batteries are slightly different in terms of their power densities and other electrical properties that govern how much energy they store and how easily it flows. We could never be certain that the charge given was the same, even when we meant it to be. Also, batteries with only a little bit of charge tend to discharge erratically. Stymied by these physical realities in our water world, we abandoned the solar ship.

This left us with our old indirect fitness nemesis: calculate a host of performance metrics and put them together to yield a single number, the individual's evolutionary fitness. With the addition of predator-avoidance to the mix, we obviously needed to shake things up a bit. Based on what we'd learned about Tadro3 in Chapter 4, we reasoned that feeding behavior was measured well by two sub-behaviors: (1) average speed throughout the trial and (2) average distance from the light.[14] We took Rob's point about acceleration to heart and added three additional sub-behaviors we thought were critical for avoiding predators: (3) peak acceleration during an escape, (4) number of escapes, and (5) average distance from the predator.

Rob had also made the point that we needed to scale each of these sub-behaviors by how much they varied among the individuals we were testing. For those of you who know and care about statistics, he suggested we use what's called a z-score. In our case, we start by taking the difference between, say, individual 5's peak acceleration and the average peak acceleration of all six of the Tadro4 individuals in that generation. That difference is then scaled by dividing by the standard deviation of the peak acceleration for all individuals in that population. For the predator-prey world, then, an individual's evolutionary fitness is the sum of the z-scores from the five sub-behaviors.

In that explanation of z-scores you can see that we also addressed another of Rob's criticisms of Tadro3: we expanded the population size from three to six for Tadro4. When we told this to Rob, he burst out laughing: "Wow! Six individuals. That's a really big population, John." Where would we be without friends to point out sarcastically the ridiculous? But here, again, the ugly anterior extremity of actually running the experiment rears up. When we double the population size, we double the amount of tail building and video analysis that we have to do. We thought we'd give six a try and see if we could survive. We did, and as you'll see, we're darn proud of all of that work!

The work that we do is also multiplied by the number of generations that we run. Rob wanted more generations because he knew that evolutionary change is usually slow and gradual. Also, he had seen that our populations of digi-Tad3s needed, at a minimum, one hundred generations to find the equilibrium of optimal stiffness. But given that we were already doubling the work by doubling the population size, the best we could promise here was to try to make it to ten generations.

CAN'T PLAY WITHOUT VERTEBRAE

The final criticism—no vertebrae—was one that we all knew from the start of the whole Evolvabot project. When Joe Schumacher had made Tadro2 he had used a long eraser as the notochord and snapped tube clamps around it to simulate rigid vertebrae. This was a great solution at the time. When we designed Tadro3 we had focused on building a biomimetic axial skeleton, in part to deal with the kind of criticism that Bob Full had about our mechanistic accuracy at the level of the skeleton. Because the notochord of the biomimetic skeleton was built from molecular collagen, the stuff of real animal connective tissues, we were pleased as punch on the accuracy side of things. However, as soon as we tried to put Joe's tube clamps on the biomimetic notochord, we destroyed it. The gelatin just was too brittle to withstand much in the way of squeezing or clamping; it would form small cracks that would then propagate into wholesale and catastrophic notochordal failure during swimming.

"Damn the torpedoes—full speed ahead!"[15] We had to have vertebrae. In Chapter 3 we talked about why building a vertebral column was a challenge: we have to attach dry and rigid structures, the vertebrae, to wet and flexible ones, the intervertebral joints. The composite assembly of alternating vertebrae and joints needs to be mechanically robust enough to be used as the propulsive tail of a Tadro. In addition, the joints, where the bending occurs, have to be of a biologically realistic stiffness. At the time of Tadro3 we'd been unable to meet this design specification, so we compromised by building the continuous hydrogels of different material stiffnesses.

Our failures with Tadro3 taught us that we needed to come up with some new tricks for Tadro4. Because we knew that working in isolation can sometimes produce creative and unexpected results, we decided to split our multi-institutional team in two.[16] Our first marine platoon was led by Tom Koob at the Shriners Hospital for Children in Tampa, Florida. Working with Adam Summers, associate professor and associate director of the Friday Harbor Laboratories at the University of Washington, and Adam's PhD student, Justin Schaefer, Tom's team took the high road. First, they created beautiful double cup–shaped vertebrae, just like the kinds you find in sharks, in a software program engineers use called SolidWorks. They then used a rapid prototyping machine to convert the three-dimensional software objects into 3-D physical objects (Figure 6.2).

A single vertebra is composed of a number of structures: a vertebral centrum, the cylinders shown in Figure 6.2 that form a chain of bones separated by the intervertebral joints; a neural arch, a rigid

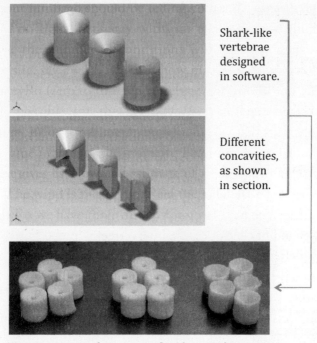

Shark-like vertebrae designed in software.

Different concavities, as shown in section.

Biomimetic vertebrae, created with a rapid prototyper.

FIGURE 6.2. *Biomimetic vertebral centra.*   Based on the structure of vertebrae seen in sharks, these biomimetic centra were designed by Adam and Justin in a software called SolidWorks. Centra varied in terms of the angle of the cup-like surfaces that attach to the flexible material of the intervertebral joint. Compared to the joints, the centra are rigid. The rigidity is created by adding cyanoacrylate glue to the powder matrix out of which the vertebrae are made. The centra are not arranged here as they will be in the biomimetic vertebral column.

structure running along the top of the centra that forms a c-shaped covering over the nervous system's spinal cord (note: the spinal cord does *not* run through the intracentral canal, the hole that runs through the centrum shown in Figure 6.2); sometimes a neural spine, a spike of bone that shoots up off the top of the neural arch; a hemal arch, the mirror opposite of the neural spine, covering the major veins and arteries that run under the centra posterior to the anus; some-times a hemal spine, a spike of bone that shoots down from the bot-tom of the hemal arch.

Biomimetic vertebral column, model 1: ligament-linked

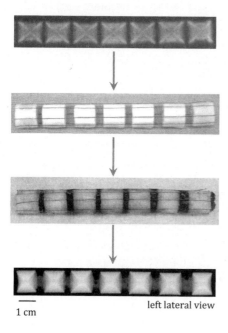

Biological target:
vertebral column of the
bonnethead shark,
x-ray showing internal concave
shape of vertebral centra.

Biomimetic vertebral column:
stage 1 — link centra
with 8 horse hairs acting as
ligaments to link vertebrae.

Biomimetic vertebral column:
stage 2 — inject gelatin into
intervertebral spaces.
stage 3 — gel then cross-link.

Biomimetic vertebral column:
x-ray showing internal structure.

left lateral view

1 cm

Features that we can vary and evolve:
1. length of centrum
2. angle of the concave joint surface
3. length of the intervertebral joint

FIGURE 6.3. *Biomimetic vertebral column, model 1.* Using the vertebral column of the bonnethead shark, *Sphyrna tiburo*, as our biological target, Justin and Tom assembled the realistically shaped vertebrae (see Figure 6.2) into a column. Horse hairs were glued to the outsides of the centra to hold the column together in a manner similar to real intervertebral ligaments. Gelatin was injected in between to create the intervertebral joints. To keep the gelatin stable and to adjust its stiffness, it was chemically cross-linked, a procedure that preserves this soft and wet material.

The exact structure of a vertebra depends on which species you are looking at and where along the vertebral column you happen to be looking. Because we were using the caudal vertebrae of sharks as our biological target (Figures 6.2 and 6.3), let me describe them.[17] Compared to those of bony fishes, the vertebrae of sharks are relatively

Vertebral column of a shark, *Squalus acanthias*, freshly dissected.
The neural and hemal arches can't be seen here. Left lateral view.

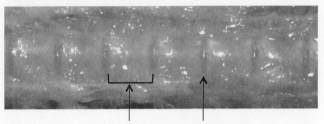

centrum   intervertebral joint

Vertebral column of a shark, *Mustelus canis*, dried.
Left lateral view.

neural
arch

hemal
arch

centrum   intervertebral joint

FIGURE 6.4. *The vertebral column of sharks.* These portions of the vertebral column are from the region between the end of the abdominal cavity and the beginning of the caudal fin.

simple, lacking neural and hemal spines (Figure 6.4). The neural and hemal arches are not fused to the centra, and those arches form their own small-diameter columns that span the intervertebral joints.

To make a vertebral column Justin and Tom figured out how to sew the vertebrae together using long horse hairs glued to the perimeter of each element. Because the hairs serve the same function as the intervertebral ligaments of real vertebral columns, we called this model 1 the ligament-linked artificial vertebral column (Figure 6.3). Between the vertebrae they injected gelatin, like the marshmallow in the middle of a camper's s'more. Once the gelatin firmed up, the whole column was bathed in a chemical fixative. This fixative cross-linked the gelatin, making the molecular collagen into a lattice that was both stiffer than the raw gelatin and resistant to degradation. This may sound familiar: the hydrogels that we made from gelatin and cross-linked were our artificial notochords that functioned as the axial skeleton in Tadro3.

Our second marine platoon worked at Vassar and was led by Kira Irving, a major in our neuroscience and behavior program at the time. Kira had been part of the Tadro3 team, along with Keon Combie, a major in biochemistry, and Virginia Engel and Gianna McArthur, both majors in biology. This group also got help from Kurt Bantilan, another major in neuroscience and behavior, and Carl Bertsche, Vassar's resident expert in machining.

In response to the problem of holding together a chain of rigid and flexible elements, Kira's team came up with a solution quite different from that of Tom's (Figure 6.5). Instead of using horse hairs as ligaments, they used thin plastic coffee stirrers as neural and hemal arches running along the top and bottom of the column, spanning each joint and preventing dislocation. This was a shark-like solution (see Figure 6.4), and it gave us the ability to explore something unusual: having very long intervertebral joints. We call this model 2 the arch-linked artificial vertebral column.

Both model 1 and model 2 gave us three structures to evolve: (1) the length of the centrum, (2) the angle of the concave joint surface on the centrum, and (3) the length of the intervertebral joint. In this respect, they were equivalent models. Where they differed dramatically, however, was in how they operated mechanically when we bent them. Model 1, the ligament-linked vertebral column, appeared to be dominated by the mechanical properties of the horse hairs. Doug Pringle, a gifted mechanical engineer in Tom's lab, helped Justin perform bending tests on the biomimetic vertebral columns. Instead of stretching the horse hairs on one side (the convex side) and squishing the joint material on the other side (concave side) during bending, they noticed that the horse hairs were stiff enough that they weren't allowing much stretching. As a result, the model 1 column bent by compressing one

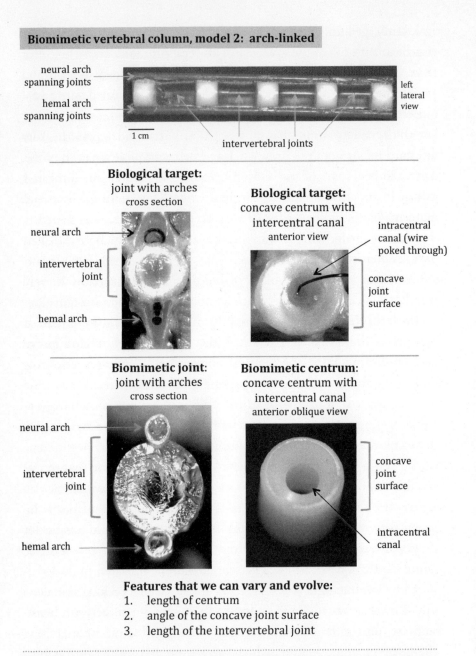

Biomimetic vertebral column, model 2: arch-linked

neural arch spanning joints

hemal arch spanning joints

left lateral view

1 cm

intervertebral joints

Biological target:
joint with arches
cross section

neural arch

intervertebral joint

hemal arch

Biological target:
concave centrum with
intercentral canal
anterior view

intracentral canal (wire poked through)

concave joint surface

Biomimetic joint:
joint with arches
cross section

neural arch

intervertebral joint

hemal arch

Biomimetic centrum:
concave centrum with
intercentral canal
anterior oblique view

concave joint surface

intracentral canal

Features that we can vary and evolve:
1. length of centrum
2. angle of the concave joint surface
3. length of the intervertebral joint

FIGURE 6.5. *Biomimetic vertebral column, model 2.* Using the vertebral columns of sharks as our biological target, Kira worked with Virginia, Gianna, and Keon to design a column that was stabilized by neural and hemal arches that spanned the vertebral column. Carl built molds that allowed our team to glue coffee stirrers onto centra in a repeatable process. The molds were then injected with gelatin, which, once gelled, was chemically cross-linked.

side, buckling locally at each joint. If you can picture this geometry in your head, then if one side of the column compresses while the other stays the same length, then the whole structure bends and shortens.

Shortening of the column didn't happen in model 2, the arch-linked vertebral column, because it used the coffee stirrers to hold the length of the column constant along its midline. During bending we saw compression of the concave side of the joint and elongation of the convex side. This looked good—at first. However, on the elongated side of the bend, we could sometimes see a separation of the hydrogel material from the face of the rigid vertebra. When this occurred it meant that the bending properties of the joint were being controlled only by the hydrogel on the compression side and the neural and hemal spines along the midline. To keep the joint attached to the vertebra, we found that we could put in a little bit of cyanoacrylate glue.

Both models of the vertebral column are beautiful examples of biomimetic design (not that I'm biased or anything). They capture the composite nature of real vertebral columns, creating a serial column of rigid and flexible material.[18] They both use a collagen-derived hydrogel for the viscoelastic intervertebral joint. They both have vertebral centra with the cup-shaped joint surfaces that we see in sharks. In addition, even though the two kinds of columns bend in different ways, their stiffness is in the same range as the stiffness we find in the vertebral columns of sharks. In order to fine-tune our biomimetic designs, we continue to explore the mechanical behavior of both the biomimetic vertebral columns and the shark vertebral columns under the expert guidance of Dr. Marianne Porter, a postdoctoral researcher in my laboratory.

The biomimetic design from Tom's lab was the preliminary winner. We thought we could solve the problem of the too-stiff horse hairs by using either fewer hairs or a different kind of material for a ligament. In addition, the arch-linked design, by virtue of its arches, had bending stiffness even when no joint was present. As with the horse hairs, we could solve that problem by reducing the stiffness of the arches. With the arch design we were also having some problems making the vertebrae and, as mentioned, keeping the joint attached to the vertebrae. The rapid prototyping, because it was computer controlled, made centra more repeatably than our hand-milled process. In addition, the horse hairs, by enforcing bending by buckling, never let the vertebrae unleash the joint material.

Thus, we began with the ligament-linked model 1. Because we were performing the robotic experiments at Vassar, we needed to train our students how to make the model 1 biomimetic vertebral columns. Because Tom and I had been meeting for research every fall at the Mount Desert Island Biological Laboratory in Salisbury Cove, Maine, we decided to use our time there to transfer the production technology. Keon and Virginia came with me to learn the task. Keon, always very keen on new techniques, volunteered to be the first trainee. Despite Tom's patience and Keon's prowess as an experimentalist, the manufacturing of the ligament-linked columns was proving to be slow at best, and a mess at worst. Imagine having to align seven small objects, keep them equally spaced, and then glue long fibers to their outsides. We built little rigs to help align and hold all the parts, but still, on most days we found that Keon soon became one with the vertebral column. Virginia and I were even worse. Our collective failure made us appreciate the abilities of Justin, who made the series of original model 1 vertebral columns down in Tom's lab in Florida. However, Justin, working on his own PhD project, was unavailable for full-time work in our vertebral column factory at Vassar.

When we returned to Vassar bearing the bad news about model 1, Kira resolved to find a better way to build model 2. Rather than linking with the stiff coffee stirrers, she designed a method to hang the vertebrae, like a string of pearls, into a mold and pour gelatin around the whole construct. The gelatin encased the vertebral column, forming a sort of pig-in-a-blanket. I loved this idea because the blanket encasement was the beginning of a body, a structure that we had blithely ignored in our previous attempts to build a vertebrated tail. With a bit more work, we thought that this new vertebral column + body design would work.

When Kira, Keon, Kurt, Gianna, Virginia, and I contemplated honing the new pig-in-a-blanket design, we had an epiphany: stop! We were spending all of our time making biomimetic vertebral columns and none of it evolving robots. The lack of a working Tadro4 system was particularly poignant because we had just welcomed into the lab a talented robot engineer, Nicole Doorly, a major in cognitive science. In order not to lose her and to move our research program along, we had to settle quickly on a vertebral column design that, in our unsteady hands, we could scale up in order to produce reliable and custom columns at the rate that we needed.

Compromise. Everybody hates that word. A compromise, it is rumored, is guaranteed to satisfy no one. No one on Team Tadro was going to be satisfied with a compromise because we'd put so much time into the design and prototyping of the fancy models 1 and 2. We were scared of the *bête noire* of compromise under the bed, knowing all the work that we were about to discard. Enter sandman: "Take my hand. We're off to Never-Never Land."[19]

In the new predator-prey land of Tadro4, the biomimetic-vertebral-column compromise consists of an artificial notochord, borrowed from Tadro3, outfitted with a series of rings forming vertebrae (Figure 6.6). This model 3, which we call, surprisingly, the ring centra vertebral column, has a number of advantages, he says bravely, over the previous vertebral column models: (1) it doesn't need to be held together by ligaments or neural arches, (2) it has fewer parts than either model 1 or 2, and (3) it can be assembled more quickly, in about five minutes (compared to thirty minutes) once all of the parts are present.

What we lose with model 3, though, is the ability to evolve the shape of the centra because the ring centra have no cup-like joint faces.[20] With that simplification, we also lose the close approximation of shape to the vertebral columns of sharks. And we're not done yet. We further simplified manufacturing, keeping the overall length of the vertebral column constant.

Simplification begets simplification. A constant column length plus unchanging centra meant that only the length of the vertebral

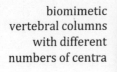

Biomimetic vertebral column, model 3: ring centra

biomimetic
vertebral columns
with different
numbers of centra

cross-linked
hydrogels
(notochords)

biomimetic vertebral column as part of the propulsive tail

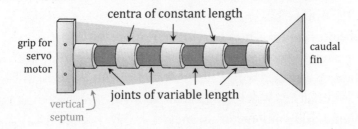

centra of constant length

grip for
servo
motor

caudal
fin

joints of variable length

vertical
septum

Feature that we can vary and evolve
length of the intervertebral joint
(vary the number of vertebrae)

FIGURE 6.6. *Biomimetic vertebral column, model 3.* Selected for use with the Tadro4 PreyRo, this model was created by the team working on model 2 and new-comers to the lab, including Hannah Rosenblum, Elise Stickles, Hassan Sakhtah, and Andres Gutierrez. Featuring ring centra slipped on and glued to a continuous hydro-gel (top image), this model is a compromise between having vertebrae and being able to build many columns quickly and in a repeatable manner.

joint could change as the number of vertebrae did. Simple! Increase the number of vertebra, and the amount of intervertebral joint available for bending decreases, thereby stiffening the vertebral column.

The full manufacturing process for model 3 involved making a slew of hydrogels, cross-linking them all in the same way so as to produce artificial notochords of similar material properties, and then gluing ring centra to each notochord to create an artificial vertebral column (Figure 6.6). This process was scaled up to production level for the game of life with Tadro4, under Gianna's supervision, with Hannah Rosenblum, Hassan Sakhtah, Elise Stickles, and Andres Gutierrez operating our assembly line.

DO VERTEBRATE CHARACTERS
EVOLVE INDEPENDENTLY OR IN CONCERT?

Because we distilled the evolution of the vertebral column down to a single trait—number of vertebrae—we had a system that allowed us to test our next biological hypothesis: selection for enhanced feeding behavior *and* predator avoidance drove the evolution of vertebrae. Testing this hypothesis also relies on the design of the stuff attached to the vertebral column, which I haven't told you about yet. We need the body with sensors to track light and predators, a microcontroller to compute the two-layer subsumption neural system, and motors to flap and turn the tail. All of this comes together in Tadro4 (Figure 6.7).

Tadro4 is really two different kinds of robot: an Evolvabot that we call "PreyRo" (*Prey* and *Ro*bot) and a nonevolving robot predator, "Tadiator" (*Tad*pole and Glad*iator*). When PreyRo and Tadiator interact in a water world with a light source, this is the Tadro4 predator-prey world (Figure 6.8). The fact that Tadiator doesn't evolve doesn't worry us, by the way. In biological predator-prey systems, predators are often much longer lived than their prey. Not that an evolving predator wouldn't be interesting! But we've got to leave something to do in the Tadro5 world.

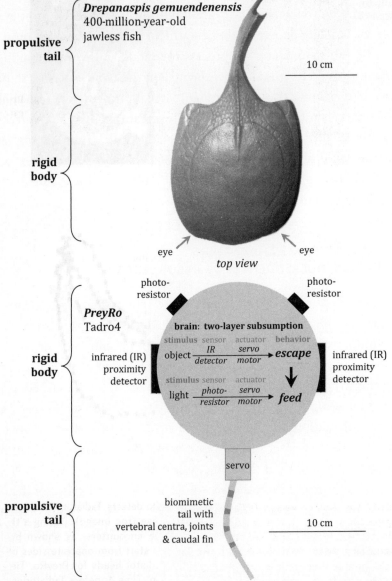

Drepanaspis gemuendenensis
400-million-year-old
jawless fish

propulsive tail

10 cm

rigid body

eye   eye

top view

PreyRo
Tadro4

photo-resistor   photo-resistor

brain: two-layer subsumption

rigid body

infrared (IR) proximity detector

infrared (IR) proximity detector

stimulus   sensor   actuator   behavior

object — $\dfrac{IR}{detector}$ — $\dfrac{servo}{motor}$ → ***escape***

stimulus   sensor   actuator

light — $\dfrac{photo\text{-}}{resistor}$ — $\dfrac{servo}{motor}$ → ***feed***

servo

propulsive tail

biomimetic tail with vertebral centra, joints & caudal fin

10 cm

FIGURE 6.7. *PreyRo is a Tadro4 Evolvabot.* PreyRo is modeled after the early vertebrate fish, *Drepanaspis*, here shown in a photograph I took of Louis Ferragalio's 1953 model at the American Museum of Natural History (specimen 8462). Both target and model share the following features: rigid, nearly circular body; body flattened in the dorso-ventral direction, like a pancake; a short propulsive tail; and paired eyes. Other features of PreyRo are based on what we know about living fishes: a lateral line for predator detection (IR proximity detectors); a two-layer subsumption neural architecture; a vertebral column with shark-like vertebral centra.

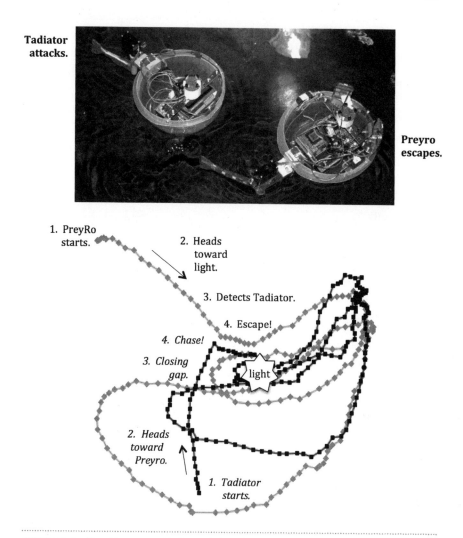

FIGURE 6.8. *PreyRo versus Tadiator.* When PreyRo detects Tadiator, the nonevolving predator robot, it initiates an escape response (top image). During a three-minute trial PreyRo and Tadiator have multiple encounters, as shown by the overlapping paths. At time 1 PreyRo and Tadiator start from opposite sides of the tank. Time 2: PreyRo heads for the light and Tadiator heads for PreyRo. Time 3: PreyRo detects Tadiator as Tadiator closes the gap. Time 4: before Tadiator gets to PreyRo, PreyRo turns rapidly away and escapes. That first close encounter took just twenty seconds. Each point on the paths represents the position of PreyRo or Tadiator each second.

PreyRo is modeled after a species of Paleozoic fish called *Drepanaspis gemeundenensis* (Figure 6.7). *Drepanaspis*, a jawless marine fish living 400 million years ago, swam using a short flexible tail and a rigid, flattened disk of a body, which lacked paired fins of any sort. Preserved in the bones that form its rigid body disk is the evidence of its sensory systems: a pair of small and widely spaced eyes and a lateral line system.[21] The flattened shape of the body of *Drepanaspis* is similar to that of living skates, stingrays, and electric rays, most of whom spend time on and in the ocean floor, feeding, burrowing, and resting. As Marianne pointed out, electric rays may be the most similar living species to *Drepanaspis* in terms of locomotor function if not ancestry because neither has or had the ability to use its body disk for propulsion. They generate thrust with a short tail that makes the whole animal look like a pancake propelled by a headless fish pushing, tugboat-like, from the rear. What we know about electric rays is that they are able to swim up in the water column. We suspect that *Drepanaspis* had the same ability.

Critters with hard shells, like *Drepanaspis* or any number of mollusks during those ancient times, appear to have been under intense selection pressure from predators, evolving tough armor and sea bottom–loving habits in response.[22] But even shells and armor could be crushed by the giant jawed fishes of the day, like *Dunkleosteus*, who had the prerequisite size, skeletal structure, and muscle strength to do the job, as shown by Mark Westneat, curator of zoology at the Field Museum of Natural History and director of the Biodiversity Synthesis Center.[23] So if you were smart, in an evolutionary sense, you, as a jawless fish and potential meal, couldn't rely on armor alone. Best not even to test your armor in the first place. Run!

Cowards get a bad rap. But dying sucks, remember? Cowards survive, for the moment, by running away. Running or swimming or flying away turns out to be the nearly universal response of animals to danger. Only those glued to a rock, hiding in a burrow, blessed with camouflage, or hormone-crazed in mating season don't flee to escape from danger. As we've seen, fish rely on the lateral line to detect dangers, and we built one using an infrared (IR) proximity detector, a small device that emits an IR pulse and uses the time it takes for the pulse to bounce back off an object to calculate the distance to the object. The mechanism is quite different, but the function is the same.[24]

Nicole, our chief engineer on the Tadro4 project, put an IR sensor on each side of PreyRo along with the two photoresistors serving as eyespots (Figure 6.7). The onboard microcontroller continuously samples both kinds of paired sense organs. If you remember the subsumption architecture we described in Chapter 5, then you can probably see the solution. At the default, or lowest level, PreyRo forages and feeds, using the difference in light intensity between the two photoresistors to calculate the direction to the light source. PreyRo is constantly making adjustments in its heading to make that difference zero, where a difference of zero means that it's headed straight up the light-intensity gradient. And so PreyRo forages to feed on the light. Until, that is, the escape behavior overrides the forage-and-feed behavioral module.

The escape behavior is triggered when either the left or right IR proximity sensor detects an object within a preset threshold distance. If the left sensor triggers, then PreyRo interrupts the forage-and-flee behavior and initiates a fixed turning maneuver to move quickly to the right. The opposite is true if the right sensor triggers. Because this "predator detection threshold" can be altered in the programming of the microcontroller, we could evolve it.

This sensory character gives us a crude way to test Rob's prediction that the evolution of the vertebrate nervous system—characterized in part by its paired sensory systems—has to evolve first in order to permit vertebrae to evolve. In other words, we predict that the evolution of vertebrae is contingent upon the prior evolution of, in this specific case, the sensitivity of the predator-detection system. This makes functional sense: why would the enhanced propulsion that vertebrae bring ever evolve without some means of detecting when it's time to use it? The only way you know is to have a sensory system capable of detecting the predators. And although the eyes you use for foraging and feeding give you some predator-detection capabilities, you can't see at night or in the dark of the deep. A lateral line, however, works anytime and anywhere. But just because a pattern makes functional sense to us doesn't have any bearing on how and why that pattern actually occurred.

If the proposed functional codependence were true, I'd be sorely tempted to call the pattern "contingent-sequential evolution." Sorry for the mouth-full phrase, but one can't be too careful when speaking of evolutionary phenomena. The contingency refers specifically to an identified causal interaction of the characters. Without a causal mechanism, we just have a correlation. Correlations may occur by accident, for no other reason than two unrelated things happen to share a pattern. However, we never want to ignore correlations because functionally codependent systems are always, in some way, correlated.

When the pattern of the evolution of two or more characters are correlated, then that pattern is called—you guessed it—"correlated evolution." When the correlation has a causal basis, it is called—you'll never guess it—"concerted evolution." Maybe you can see why, for a specific sequenced pattern of concerted evolution, I've proposed the phrase "contingent-sequential evolution." Whatever the pattern of concerted evolution, to claim that you have one requires that you show the specific functional codependence between or among the characters. Categories of potential functional mechanism include genetic, developmental, and physiological.

Whereas concerted evolution means that two or more characters evolved because of their interactions, when no interactions are present we call that a pattern of "mosaic evolution." Mosaic evolution was introduced in Chapter 2 so that we could make the important point that species are not "primitive" or "derived" but are, instead, mosaics of both ancestral and derived characters. Mosaic evolution is a fact of life, but it's not the only fact of life. Concerted evolution is also a fact. If we look at enough characters, we will see both kinds of character evolution in any species: mosaic and concerted.[25]

In addition to our predicted lateral-line → vertebrae pattern of sequential-contingent evolution, we also expect concerted evolution within the propulsion system itself. A character of all fishes, extinct and living, that varies like crazy is the shape of the caudal fin. The earliest known vertebrate, *Haikouichthys*,[26] has a caudal fin that tapers to a point, like an eel. Yet our Tadro4 target, *Drepanaspis*, has a two-lobed caudal fin that splays apart, forming a sharp vertical trailing edge, or "Kutta condition," to use the hydrodynamic lingo.[27] These two kinds of tails, tapered and splayed, are just some of the kinds that we see. In terms of propulsion, what matters is the length of that trailing edge, measured as the "span" of the caudal fin. The trailing edge is where the body sheds so-called bound vorticity into the water. If we were to revisit Lighthill's Elongated Body Theory (EBT), which we used earlier to propel our digi-Tad3s, then we'd see that propulsive power generated by the fish is proportional to the square of the tail's span. That square term is huge: a slightly larger tail span should help produce much more power.

Because both the tail's span and the vertebral column are involved in generating propulsion, we predict that the two characters will show concerted evolution. Here's why, specifically. In order for the square-of-the-span magic to work, what I told you above assumes that everything else about the motion of the fish's body stays the same, including how far it moves its caudal fin side to side, what we measure as the lateral amplitude of the caudal fin. But that caudal fin amplitude will decrease when you put a bigger caudal fin on the body—it has to. It's like when you, ignoring the better judgment of your parents, used to stick your hand out the window when speeding down the highway. Palm down, hand parallel to the road? No problem. Rotate your hand ninety degrees. Boom! Your hand flies backward and you ram your arm into the window frame. Ouch. The difference is drag, which is low in the first position and high in the second position.

A tail with a larger span has more drag when it moves laterally than does a tail with a smaller span. When the small span is increased to the large, the only way to keep the amplitude of the caudal fin constant in the face of the increased drag is to generate more power internally, in the machinery that is driving the tail. Guess what? That internal machinery includes the vertebral column, which we know stores and releases elastic energy as it bends. So here's the basis of our concerted-evolution prediction: a stiffer vertebral column could compensate for the increased drag that accompanies the increased span of the caudal fin.

TESTING A TRIO OF REALLY COOL HYPOTHESES WITH TADRO4

We set out to test the hypothesis that selection for enhanced feeding performance and predator avoidance would increase the number of vertebrae. Next thing you know, we are yammering on about two other characters that we are going to evolve: the predator-detection threshold and the span of the caudal fin. We predicted that both of these other traits would evolve in concert with the number of vertebrae, with predation-detection evolving before vertebrae (the sequential-contingent pattern of concerted evolution) and the span of the caudal fin evolving at the same time (just plain old concerted evolution). The alternative hypotheses are that predation-detection may evolve at the same time as number of vertebrae (concerted evolution) or may not be correlated with the number of vertebrae (mosaic evolution). Span of the caudal fin, too, may not be correlated with the number of vertebrate (mosaic evolution).

The results of our evolutionary experiments in the predator-prey world of Tadro4 are fascinating (Figure 6.9). Our amazing Tadro4 teams met and exceeded all expectations. Led by Gianna in the summer of 2007, Hannah, Elise, Andres, and Hassan perfected the biomimetic-vertebral-column production line and conducted the first evolutionary run, which lasted for five generations before we had season-ending injuries to the robots.[28] Led by Hannah and Andres in the summer of 2008, Sonia Roberts and Jonathan Hirokawa conducted the second evolutionary run, which went for eleven generations. Because we started the two populations with the same average values for their evolving characters, the two runs are independent replications of the same experiment. It turns out that comparing the runs is critical for testing our hypotheses.

First and foremost, in both runs of the predator-prey world, the population of PreyRos quickly evolves more vertebrae, moving from the starting average of 4.5 to an average of 5.5 by the third generation. In the second run an equilibrium average of 5.7 vertebrae appears to have been reached. These early directional and positive increases, even though they are modest, provide tentative support for our big-picture hypothesis that the number of vertebrae increase when the population is under selection for enhanced performance in feeding and fleeing.

I can only say "tentatively support" because, as we talked about in Chapter 4, strictly speaking you can only falsify a hypothesis—you can't prove it. So the phrase "tentatively support" is meant to recognize (1) that we have failed to falsify and (2) that over time repeated failures to falsify will eventually lead us to conclude that the hypothesis is probably true. Caution and consideration are required.

But it's really, really hard not to get totally stoked when you see this pattern of evolutionary change. Emotionally, we—okay, I—want to yell, "Kick ass! We've proven that this specific type of selection on these fish-like autonomous agents works as we predicted!" But we mustn't do that. And we mustn't holler, "And we did it twice, you cynical bastards, and it worked both times! Evolvabots rock!" So we don't. And we won't. Ahem. What was I saying?

Right. "Dignity, always dignity."[29] Please take the time to notice in Figure 6.9 that by reaching an apparent equilibrium, the population of PreyRos creates an evolutionary pattern with their average number of vertebrae that resembles the evolution pattern that we saw in the Digi-Tad3s and their average tail stiffness (see Figure 6.1). Is this coincidence? I think not. Once again we see evidence that is consistent with the hypothesis that the stiffness of the vertebrate axial skeleton has evolved to balance the mechanical demands of maneuverability, wherein a flexible axis works best, with those of speed, whereby a stiff axis outperforms.

Because we were evolving three characters—(1) number of vertebrae, (2) predator-detection threshold, and (3) span of the caudal fin—at the same time, the patterns of evolutionary change shared or not shared between them also inform. For example, in both runs of the predator-prey world, the PreyRo population's changes in the predator-detection threshold are positively and strongly correlated with the changes in the number of vertebrae, at least over the first five generations (Figure 6.9, middle graph). This strong correlation tentatively supports the hypothesis of concerted evolution between these two characters. What we don't see is a threshold-first-then-vertebrae-next pattern—with vertebrae lagging in time—that would support the sequential-contingent pattern.

This in-phase pattern of apparent concerted evolution is evidence that a functional codependence exists between sensing a predator and moving away from it.[30] Are we surprised? No. But that's why you run

FIGURE 6.9. **(facing page)** *The evolution of vertebrae in PreyRo is directional, concerted, and mosaic.* If you look at the evolution of the average number of vertebrae, N, in the population (top graph), you'll see that in both evolutionary runs N increased over the first five generations. This is a pattern of *directional selection.* In the second run N then reached a plateau, just like we saw with the tail stiffness in the digi-Tad3s (see Figure 6.1). Points represent the populations' average, and the error bars represent the standard error of the average. Only one side of the error bar is shown, by the way, so that they don't overlap when the means are close.

The predator-detection threshold (middle graph) increased over five and then three generations in the first and second runs, respectively. In both runs predator-detection threshold is strongly and positively correlated with N over the first five generations (r values of 0.93 and 0.92, where r can vary from 1 to −1). This pattern of correlated evolution with N is consistent with a hypothesis of *concerted evolution* for these characters with respect to each other.

The span of the caudal fin (bottom graph) shows different patterns in both runs, with an initial decrease followed by an increase in the first run, and a longer, stronger decrease followed by a longer, stronger increase in the second run. The correlation of the span of the caudal fin with N over the first five generations is positive in the first run and negative in the second. This inconsistent pattern of correlations with N is consistent with a hypothesis of *mosaic evolution* for these two characters with respect to each other.

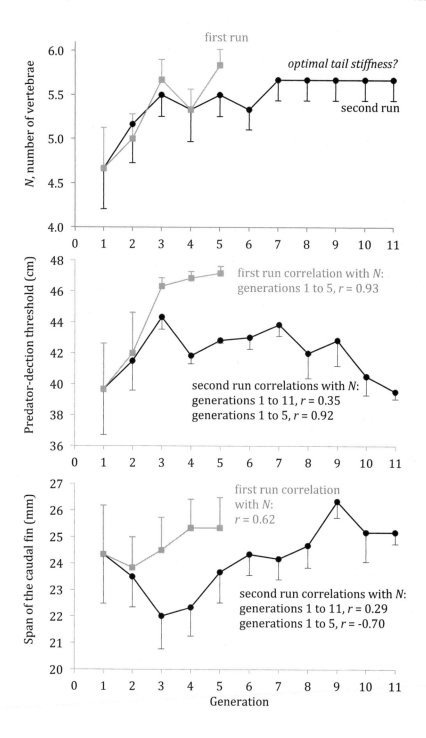

first run

optimal tail stiffness?

second run

first run correlation with N:
generations 1 to 5, r = 0.93

second run correlations with N:
generations 1 to 11, r = 0.35
generations 1 to 5, r = 0.92

first run correlation
with N:
r = 0.62

second run correlations with N:
generations 1 to 11, r = 0.29
generations 1 to 5, r = -0.70

N, number of vertebrae

Predator-dection threshold (cm)

Span of the caudal fin (mm)

Generation

the experiments. We aren't surprised because we knew, from Chapter 5, that tightly linked perception-action feedback loops (PAFL) create behavioral modules. What's new here is that we've shown that this particular PAFL—Escape!—is well characterized by the ability to detect predators and then flee from them. What's also very interesting is that we've connected the Escape! PAFL to the evolution of vertebrae. This connection, measured here as a strong and positive correlation between predator-detection threshold and the number of vertebrae, allows us to understand how selection acting on behavior changes a feature of the skeleton.

We had also predicted an in-phase pattern of concerted evolution for the character pairing of number of vertebrae and the span of the caudal fin. Here, though (Figure 6.9, bottom graph), the direction of the correlation reverses from one run to the next, at least over the first five generations. Because we don't see the same pattern in the two runs, this seems like a clear case in which we have refuted the hypothesis of concerted evolution. The default or "null" hypothesis is that the two characters show mosaic evolution, at least with respect to each other.

Mosaic evolution was a surprise. We thought for sure that these two characters, number of vertebrae and the span of the caudal fin, would show a functional connection because they are tightly connected in terms of anatomy and physiology. Again, this is why you run the experiments! This result highlights the fact that we always have to test our assumptions. I want to point out, though, that you could imagine different situations in which you might see a tight and consistent correlation between the two characters. For example, if we were just looking at swimming speed alone, devoid of any evolutionary environment, I'd predict that we'd see a relationship. So there!

BEIM SCHLAFENGEHEN

It's time to put our Tadros to bed. Tadro3, Digi-Tad3, and PreyRo (Tadro4) have done their jobs. They have evolved. Their characters—

tail stiffness and the number of vertebrae—evolved and, in so doing, tested our hypotheses about what kinds of selection pressures may have driven the evolution of early vertebrates. We've learned that whereas selection for improved feeding behavior seems unlikely to have been the sole driver of increased number of vertebrae, when coupled with fleeing from predators, it becomes a much more powerful selective pressure.

With Tadros, we've seen how to start with the simplest autonomous agents you can imagine and then add in only the smallest bit of complexity needed to make a new and/or improved model of your biological system. Moreover, the simplicity of Tadros as embodied brains gave us an opportunity to understand the physical basis of the intelligence and behavior of feeding and fleeing. Finally, Tadros serve as working examples of this special category of robots that we've come to call Evolvabots.

As the Tadros sink into slumber, we still have much to explore.

chapter 7

EVOLUTIONARY TREKKERS

REMEMBER: NO MATTER WHERE YOU GO, THERE YOU ARE.[1] That's one of the most frustrating things about the otherwise wonderful Evolvabots. They visit just a few places in a vast morphospace of evolutionary possibilities (Figure 7.1). As we've seen, a population of Tadro3s or PreyRos take but a single path out of a huge number of possible trajectories. Which path they take and how quickly they travel it depends on those three all-encompassing classes of evolutionary mechanism that we introduced in Chapter 2: selection, randomness, and history. Those mechanisms determine what we observe at any time and place in the population's evolutionary journey: a generation of individuals playing the game of life. Although the game of life is fascinating to watch, sometimes the "there" you are observing is not the "there" where you want to be.

We were curious about the travels of our two different populations of PreyRos. Why had the populations never evolved more than an average of 5.7 vertebrae? Our guess, based in part on the behavior of many populations of Digi-Tad3s, was that the PreyRos had reached an equilibrium or, possibly, an optimum number of vertebrae. But what, if anything, makes 5.7 optimum? Why not 8 or 10 vertebrae? What if PreyRos evolved 12 vertebrae: would the world end?

Why not? What if? Questions about the counterfactual dogs anyone interested in just about anything with a history, including the process of evolution. Whether we are studying the evolution of lifeforms,[2] engineering solutions, or artificial intelligence, our curiosity about what didn't happen[3] or what might have happened motivates much of what we do. A central curiosity-driven question regarding any evolutionary system is: *Why did some forms evolve while others haven't?*[4] From this question springs a host of related queries:

1. Why didn't the population evolve along a different path?
2. What if the population were to evolve again, from the same starting point? Would it evolve along the same path?[5]
3. Why haven't all imaginable forms evolved?[6]

How to proceed? Curiosity leads us to a classic forbidden-fruit conundrum, presenting us with at least three possible actions. First choice: don't bite the apple. We could simply let our findings be, content that PreyRo, in two different runs, increased the number of vertebrae under selection for enhanced feeding behavior and predator avoidance. We've learned much. Move on to the next study.

....

FIGURE 7.1. **(facing page)** *There you are.*   Evolution samples only a small portion of the morphospace, the area of all possible trait combinations. The entire second evolutionary run of PreyRo is shown here, with the points representing the average values of the population for the pair of traits that we found to be evolving in an uncorrelated, mosaic pattern, the span of the caudal fin, *b*, and the number of vertebrae, *N*. The thin vertical and horizontal gray lines intersecting each point represent the ranges of each trait for a given generation. The rectangles that those range lines touch represent the population's total phenotypic footprint, the area that the population has sampled.

Top: The population of PreyRos evolves over a very small region of morphospace in ten generations. The rectangle encloses all individual PreyRos ever evolved.

Bottom: A close-up of the evolution of the population in *b-N* morphospace shows a shrinking phenotypic footprint as selection removes variation, here represented by the ranges, from the population. Loss of variation in traits is a sign of selection.

....

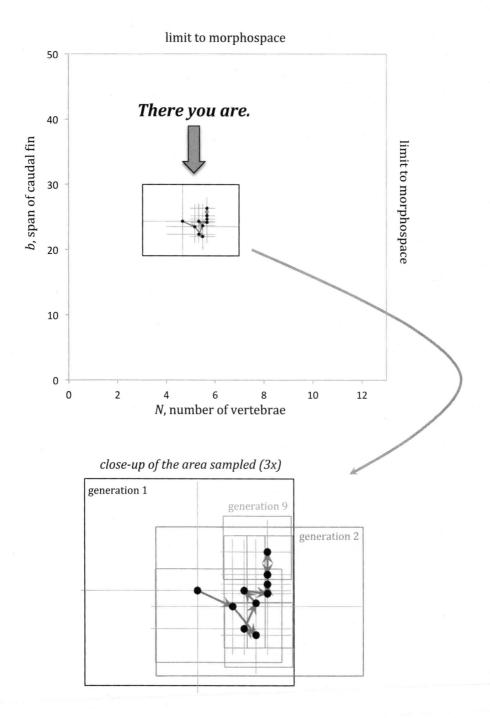

There you are.

limit to morphospace

limit to morphospace

b, span of caudal fin

N, number of vertebrae

close-up of the area sampled (3x)

generation 1

generation 9

generation 2

But no, for our curiosity remains. Remember this inscription that taunts Digory in C. S. Lewis's *The Magician's Nephew:*

> Make your choice, adventurous stranger
> Strike the bell and bide the danger
> Or wonder, till it drives you mad
> What would have followed if you had.

Madness, surely, is not something that we desire. If wonder we must, then adventure we take. Strike the bell! Here's an adventurous choice: employ "directed evolution" and then try to force the system to evolve toward a prescribed goal or along a particular path. Although this approach to engineering might seem new—it has garnered much attention recently because of its success in the synthesis of novel proteins—you could argue that directed evolution is what breeders have been doing for millennia to domesticate the likes of cattle, rice, and corn. It works, but only to a certain point and in certain cases. No matter how much I want flying cows or talking corn, it ain't gonna happen given the limits genetics and physics impose. Just because you have a target doesn't mean that you'll be able to hit it or, for that matter, that it's even hittable.

We have another choice that helps satisfy our curiosity. We can hit the target by knowing exactly where we want the robot to be in the evolutionary landscape. Unlike chemists trying to engineer enzymes of unknown structure with a targeted function, we, as mentioned in Chapter 3, take the reverse-engineering course, asking about the function of a given structure (rather than seeking a particular function). For example, we could just build a PreyRo with ten vertebrae and see how it functions. This is like dropping a paratrooper behind enemy lines. We can put individual robots at specific points on the evolutionary map and ask them to report back. These robots serve as probes, allowing us to go where no one has gone before. In homage to *Star Trek*, let's use eponymism (see Chapter 3) and call these agents Evolutionary Trekkers, or ETs for short.

As we've defined them, ETs are not Evolvabots. They don't evolve. Sorry. Just like the rules for Starfleet officers, the prime directive (General Order 1) applies, and ETs don't participate in or alter the evolutionary trajectories of the prewarp life-forms that they study or encounter. Even with this limitation, they are a powerful complement to Evolvabots. ETs are the crew members testing the functional waters, so to speak, of different bodies and brains without the historical constraint that Evolvabots drag with them. Keep in mind that ETs cannot test different selective conditions that might drive a system to evolve from one place to another on the adaptive landscape. For this reason, ETs are a separate class of robots, testing hypotheses about the *outcomes* of evolution rather than the process itself.

MAPPING THE EVOLUTIONARY LANDSCAPE

Before we explore ETs I need to clarify some terms that I've tossed around willy-nilly. The term "evolutionary landscape" is also known as a "fitness landscape" or, in the original concept Sewall Wright created, an "adaptive landscape": I use these terms interchangeably. The metaphor of a landscape gives us a way to conceptualize the hills of fitness heights and the valleys of fitness despair. Fitness is represented by contours lines on a two-dimensional map (Figure 7.2).

You've probably noticed that I didn't show you all three of PreyRo's evolving traits in Figure 7.2. That's for the simple reason that maps with more than two traits are difficult to make and interpret visually. For example, for three traits in a three-dimensional surface, you need to be able to rotate the surface so that you can view the selection vectors from different angles. On top of that, literally, you then have to somehow visually code the fitness gradients. Folks more talented than me with visual graphics can manage. But we all fall down when it comes to illustrating maps with more than three dimensions.

Even though adaptive landscapes, as a visual tool, have severe limits, at least for the two characters shown here, they can be very instructive (Figure 7.2). To wit: we had previously decided that the number of vertebrae, *N*, and the span of the caudal fin, *b*, evolved independently with respect to each other, a pattern of character interaction that we called mosaic evolution (see Chapter 6). Because this *N-b* character pair is mosaic and thus uncorrelated, we can't tell what will happen to one by simply looking at the evolutionary changes of the other. Instead, we make sense of the combined evolutionary history that they share, even if it is uncorrelated, by looking at the population's evolutionary trajectory and the adaptive peaks in the *N-b* landscape.

What we see on the map is wild (Figure 7.2). From a bird's-eye view we see multiple adaptive peaks, a chain of misty fitness mountains running from north to south. In between the peaks we find what looks like valleys and then a whole bunch of white space labeled "terra incognita." All of the white space on the map, and probably some of the gray hilly parts too, is unknown territory.

Here we return to our main problem: adaptive topography can only be mapped if the population has been in that area—to that "there"—and played the game of life. Only when each individual gets a fitness score can we then calculate the population's selection vector. For PreyRos my colleagues and I determined where the vector pointed (say, to having five vertebrae and a caudal fin span of 22.25

FIGURE 7.2. (facing page) *Mapping the adaptive landscape.*
Top: We can use two of PreyRo's evolving traits, span of the caudal fin, *b*, and number of vertebrae, *N*, to create the two-dimensional "morphospace" of the evolutionary map. The points represent the population's average values for the traits at each generation, numbered 1 to 6 for the first evolutionary run. The black arrows represent the actual evolutionary change of the population from generation to generation. The gray arrows are the selection vectors. Each selection vector has a direction and a strength, with strength represented by the length of the arrow. The random evolutionary mechanisms (mutation, mating, and genetic drift) cause the difference between the selection vector (gray arrow) and the evolutionary vector (black arrow).

Bottom: Because the selection vectors point toward a local fitness peak, they can be used to map the adaptive landscape. The points here are the same average values of the population from above (generations 1 to 5). The arrows are the same selection vectors. Adaptive peaks and ridges can be of any shape. The shape and placement of the fitness features that I've drawn here are wildly speculative, given that we have only five selection vectors. *Terra incognita* (any white area) refers to areas that are unknown and therefore unmapped. Note that the lack of selection pressure (short arrow) means that the population is on an adaptive peak.

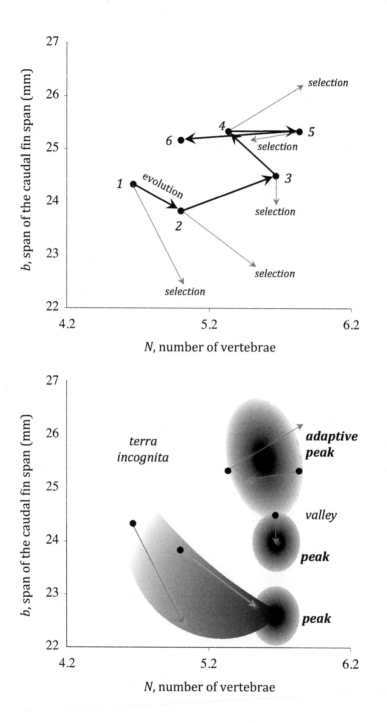

millimeters).[7] The selection vector represents the direction and magnitude of evolutionary change that selection alone would cause.

But selection does not act alone. You can see in the top diagram that those selection vectors don't predict exactly where the population moves on the landscape. Deviations between the selection vector and the actual evolutionary trajectory, which can be thought of as another vector, are caused by random processes (mutation, mating, drift). What the selection vectors do, though, is point uphill toward the closest fitness maximum or "adaptive peak."

Selection maps evolutionary *terra incognita*. With that in mind, look at our population of PreyRos in generation 3 (find the generation number in the top diagram and then look for the corresponding point in the bottom diagram, Figure 7.2). That population sits in what I've labeled as a valley, an area of low fitness compared to two or more close-by regions of higher fitness, the adaptive peaks to the north and the south.[8] The selection vector for generation 3 is small and points due south. The small magnitude of the vector means that the population is nearly sitting right on top of an adaptive peak, with just a little bit of climbing in the b-dimension to reach the local summit. The fact that this generation-3 population never summits but instead shifts off of this peak in generation 4 illustrates one of the great ironies of evolution: random factors like mutation and mating can displace a well-situated population, adaptationally speaking.

Evolutionary conditions change drastically from generation 3 to 4. The compass of selection has backed from the south to northeast. This change means that instead of finding a peak by reducing the span of the caudal fin, the population ascends a different peak now by increasing both the span and the number of vertebrae. Traversing to generation 5, the population overshoots the summit, and selection points back from whence the population came.

Mapping the adaptive N-b landscape shows dramatically how a steady selection pressure—rewarding enhanced feeding and predator avoidance—can produce a tortuous evolutionary trajectory. What we still don't know, however, is what the whole landscape looks like. How extensive is the range of adaptive peaks? Are they peaks or ridges? Do other adaptive peaks exist?

THE LIMITS OF EVOLUTIONARY BIOROBOTICS

We could overcome our evolutionary ignorance using a modified directed-evolution approach. We could plop down populations of Evolvabots somewhere, run evolutionary trials, and let selection map the local terrain. This beam-me-down-Scotty procedure would be instructive, but it would be hit or miss with respect to the whole map. We'd run the risk of doing all that time-intensive evolutionary work only to find ourselves in the middle of an extensive fitness valley.

Given enough time and money, though, you could absolutely use directed evolution to expand your map. Populations of Evolvabots placed evenly and densely throughout the entire *N-b* landscape would enable you to accomplish what my colleague Chun Wai Liew calls an exhaustive search of the parameter space. This is not realistic with physical models, but it does work fine in digital simulation, assuming your landscape has only a couple dimensions, such as number of vertebrae and tail span, and covers just a little territory. But what if you add in predator-detection threshold, shape of the vertebrae, length of the tail, activity patterns of the muscles, and all of the various neural control mechanisms? By the explosive mathematics of combinatorics, each added dimension, k, expands the possible number of combinations, n, which in our case are different kinds of genotypes or phenotypes, by the following relationship with the number of possible values or conditions, j, within each dimension:

$$n=j^k$$

Don't let this cute little equation fool you. It hides a tactical hurricane. Before the storm, the wind is fair. Let's say that we have only two dimensions—the number of vertebrae, N, and the span of the caudal fin, b—so that $k = 2$. If we allow both dimensions to have four conditions, $j = 4$, then the number of different phenotypes would be $n = 4^2$, or 16. No problem.

But your lumbago should be telling you that a storm is approaching. Hold on tight. Let's stay with our two dimensions, $k = 2$, but now make the number of conditions within each dimension a bit more realistic, say $j = 14$, which is how many vertebral states are possible in PreyRo (zero to thirteen vertebrae). Even though more conditions are possible for the span of the caudal fin (zero to fifty millimeters, in one-millimeter increments), we'll just say that both have the same j value for the moment. For just our b-N adaptive landscape, that gives us a low-end estimate of $n = 14^2 = 196$ different phenotypes. The wind is pickin' up. Reef the sails!

In PreyRo we have three dimensions, so $k = 3$. We'll stay with our conservative estimate of fourteen possible conditions in each dimension. By adding a third dimension, we now have the possibility of $n = 14^3 = 2,744$ different phenotypes. But that figure underestimates the number of conditions, because the span of the caudal fin has fifty and the predator-detection threshold has fifty (10 to 60 centimeters in 1-centimeter increments). Let's choose $j = 25$ and see what happens: $n = 25^3 = 15,625$ different phenotypes. Gale force winds! Batten down the hatches and man the bilge pumps! We are taking on water, matey.

This little exercise makes several things crystal clear. One: nautical metaphors are extremely annoying. Two: it is practically impossible, in the true sense of both words, to build and test all types of even a simple Evolvabot like PreyRo. Three: map makers wanting extensive or exhaustive maps of the adaptive landscape must resort to digital simulation because that is the only way to approach the number of trials needed in hill-searching and hill-climbing experiments. It's also worth noting that there are ways to avoid having to do an exhaustive brute-force search of an entire landscape; Chun Wai uses different kinds of evolutionary algorithms to balance the demands of finding all of the peaks and doing so in a reasonable amount of time—like weeks instead of years. He has developed a meta-algorithm that decides when to use an exploring routine to search broadly and when to switch to a focusing routine that finds local hills.[9]

Seduced by the phenomenal cosmic power[10] of digital simulations and their handlers, I can't help but wonder what in the world I was thinking. Using physically embodied robots? Evolving their biomimetic body parts? Enslaving students to work in the robot factory? Believing in autonomous behavior and situated-and-embodied intelligence? John, you dummy! Think of all the time and goodwill that you've wasted. Like it or not, digital simulation, clearly, is the way to go.

Then, a voice. I hear the Ghost of Christmas Past: "You see, John," he whispers, "if you had not become obsessed with physically embodied robots, your life and the lives of those around you would have been much different. It would've been better."[11]

Yes, I think, it would've been different . . . better. I could've explored the entire adaptive landscape of early vertebrates in the wink of an eye using digital simulation. My students and I could've moved beyond vertebrae. We might even have explored why fish evolved paired appendages, swim bladders, different body shapes, and the ability to live on land.

"John," says the ghost, this time louder, "You must change your methodology. There is still time. Join your friends in the land of digital simulation, and you will come to understand why computational biology is *de rigueur*." At the unexpected use of French, I turn, expecting to see my Gallic tormenter, hoping to plead for the opportunity to retool, perhaps in a year-long sabbatical at a stylish Parisian university or, if that's not possible, in an intensive summer course at a marine laboratory on the Mediterranean. But I see no one.

Without meaning to, I say aloud, "I guess you're right." Now I hear the ghost smile, as if that's possible, and I see him, or at least his face, beaming. "Yes, John, I am right. And you are right to change your ways while you still can, for the sake of yourself and for those around you. Now that you have seen what has been and what might be, I take my leave."

A chilly breeze rustles the Post-its on my bulletin board as I hear the ghost ask a parting question, tossed casually, as if by a long-lost colleague walking away down the hall: "One last thing, John, that always puzzled me: what do robots have to do with biology?" At last recognizing the trick well played, I freeze, caught between my self-loathing for willfully betraying embodied robots and my embarrassment for falling prey to the old Ghost-of-Christmas-Past ploy. Cue Marlon Brando as the voice-over narrator: "Horror . . . Horror has a face . . . and you must make a friend of horror."[12] Never!

Let us unmask the horror. The face revealed: the why-robots question. Aha! We have faced you before, foul query, back in Chapter 1. But we are different now—stronger. We have data. We can wrestle with you once more, emboldened now by our experience and knowledge. And we know much.

We know that the process that we've dubbed "evolutionary biorobotics" works. We know that we can design and build autonomous Evolvabots that represent and hence model extinct and living animals. We can let a population of Evolvabots loose in a simplified world, and that population will evolve under the combined effects of history, randomness, and selection. We know that we can use Evolvabots to test hypotheses about the evolution of the traits of early vertebrates. And we know that by virtue of their explicit simplicity, Evolvabots allow us to witness, interpret, and understand puzzling evolutionary patterns.

But is that enough? Wouldn't we learn the same thing—and learn it faster—from digital robots? No, no, no, whispering ghost, leave us be! The problem and the difference is physical. We don't simulate the physical world—we live it. Remember Chapter 1? I'll recap the reasons for building physically embodied robots and add, in italics, what we've learned since we first set eyes on this list.

With physically embodied robots built to model animals (what Webb calls biorobots):

1. You can't violate the laws of physics . . . *because the robots are enacting, not modeling, the laws of physics.*
2. You can build a simplified version of an animal . . . *using the KISS principle, the engineer's secret code, and Webb's modeling dimensions as guidelines.*

3. You can change the size of the animal . . . *to suit the needs of your experiment or match the physical situation of the targeted system.*

4. You can isolate and change single parts, keeping all else constant . . . *giving you a decent chance to understand the behavioral complexities that even simple agents produce.*

5. You can reconstruct extinct animals and some of their behaviors . . . *if you know enough about the anatomy and physiology of the targets and the environments in which they lived.*

6. You can create animal behavior from the interaction of the agent and the world . . . *without needing to code "behavior" into the "brain" because behavior is the dynamic spatiotemporal event that occurs when an autonomous agent operates in an ongoing perception-action feedback loop with its world.*

7. You can test hypotheses about how animals function in terms of biomechanics, behavior, and evolution . . . *if and only if your embodied robot is carefully designed to represent explicit features of your biological system.*

Phew. One thing that we've learned for sure: verbosity. More importantly: a biorobot that is embodied and situated is a physically instantiated simulation, a representation of a biological target, a model. But that's not all. An embodied biorobot is also a physical thing in and of itself. You can't take that away from me, or the robot. Even if someone, like one of our intellectual predators from Chapter 6, decides that your Evolvabot is a horrible model of an evolving fish, that Evolvabot is still, undeniably, a physical, material entity. Looks like a material entity. Feels like a material entity. Tastes like a material entity. Anyone for a bowl of material entity? Yes, please. I never eat anything but. Yum.

It's at this stage that folks like me, working with physically embodied robots, like to claim that the digitally simulated robots, the binary things on the computer, aren't real. My robots are real. It's those digital simulations that aren't. I'm okay, you're a fake.

However, I'm not going to say that, even though I just did (but I didn't mean it—paradox alert!). What I'm going to say instead, because it's a more accurate reflection of reality, is that digital simulations do indeed have a physical reality: electrons, within silicon dioxide microcircuits, by virtue of their controlled movements, carry out a series of Boolean logic functions that, in aggregate, represent the manipulation of symbols defined by a human as part of an algorithm. Those electrons interact with a world of other electrons as well as the constraints and channels of their semiconductive silicon environment. The electrons are not spirits in the material world. They have mass, charge, and velocity. They behave in the same way that embodied robots do—governed by the laws of physics—when they interact with the world. So to say that those electrons aren't real and can't behave is false.

Then why all the fuss? What's the difference between an embodied robot as a model simulation and a digital robot as a model simulation? Barbara Webb, the creator of the field of biorobotics (see Chapter 1), makes the distinction between modeling in software and hardware: "The most distinctive feature of the biorobotics approach is the use of hardware to model biological mechanisms."[13] Webb elaborates: "a more fundamental argument for using physical models is that an essential part of the problem of understanding behaviour is understanding the environmental conditions under which it must be performed."[14]

Now we are onto something. The difference is not physical simulation versus nonphysical simulation. It's not materialism versus substance dualism (see Chapter 5). The difference is how we model the behavior. Do we create the behavior by representing the interactions of agent and environment algorithmically, mathematically? Or do we create the behavior by *not* representing the interactions at all but instead letting them "just happen"? When behavior just happens, we remove a layer of the simulation, a layer of representation that, when present, increases the conceptual distance between the target and the model (Figure 7.3).

Let me put it this way: if behavior is the physical interaction of—or feedback between, if you prefer—a physical agent and a physical world, then that behavior can be modeled in mathematical representations or not modeled. This makes me think that Rodney Brooks

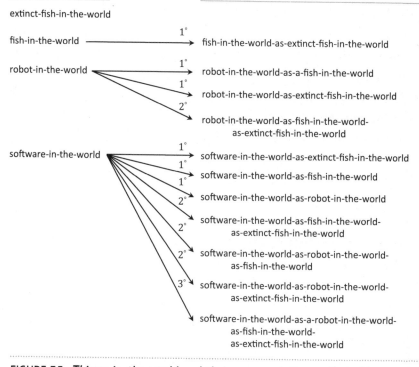

the thing-in-the-world
(not a representation)

representation...
of the thing-in-the-world (1°) or
of the representation of the thing-in-the-world (2°) or
of the representation of the representation of the
thing in the world (3°)

extinct-fish-in-the-world

fish-in-the-world ⟶ 1° fish-in-the-world-as-extinct-fish-in-the-world

robot-in-the-world 1° robot-in-the-world-as-a-fish-in-the-world
1° robot-in-the-world-as-extinct-fish-in-the-world
2° robot-in-the-world-as-fish-in-the-world-
as-extinct-fish-in-the-world

software-in-the-world 1° software-in-the-world-as-extinct-fish-in-the-world
1° software-in-the-world-as-fish-in-the-world
1° software-in-the-world-as-robot-in-the-world
2° software-in-the-world-as-fish-in-the-world-
as-extinct-fish-in-the-world
2° software-in-the-world-as-robot-in-the-world-
as-fish-in-the-world
2° software-in-the-world-as-robot-in-the-world-
as-extinct-fish-in-the-world
3° software-in-the-world-as-a-robot-in-the-world-
as-fish-in-the-world-
as-extinct-fish-in-the-world

FIGURE 7.3. *Things-in-the-world and their representations.* Each thing-in-the-world can, if carefully designed by a human, represent other things-in-the-world. The great power of software is that it can represent any thing-in-the-world, even representations of things-in-the-world. The same can't be said for fish: you would never argue, I hope, that fish-in-the-world represent software-in-the-world. But because representation depends on the intent of the human experimenter, folks can and do argue that they can use fish-in-the-world to represent extinct-fish-in-the-world (primary representation, 1°). We built Tadro3s to represent the tadpole larvae of tunicates that, in turn, we selected to represent early chordate ancestors of vertebrates (secondary representation, 2°, of chordate ancestors by Tadro3s). When we created digi-Tad3s as representations of Tadro3, we created an additional layer of representational distance from the target (tertiary representation, 3°, of chordate ancestors by digi-Tad3s).

A robot-in-the-world and software-in-the-world can both be built as primary representations of an extinct-fish-in-the-world. In that sense they are equivalent as model simulations. What differs is how they represent behavior. As part of its representation of an extinct-fish-in-the-world, software-in-the-world must represent the physical interactions of the agent and its environment. This is a hidden or implicit level of representation (let's call it 0°) that increases the conceptual "distance" between the target and the model that represents it.

was pulling our collective leg when he said, "The world is its own best model." Here's the paradox: the world is not a model; it is simply the world itself. We only make the world into a model when we force it to represent something else. This, then, is the Zen of Physically Embodied Biorobots.

As physically embodied biorobots, we've already established that ETs aren't Evolvabots: they don't evolve and, hence, they can't directly test hypotheses of evolutionary process. However, as suggested by our list of the seven reasons to use embodied robots (page 178), ETs can test hypotheses about how extinct animals functioned and behaved. Because we've taken a page from cognitive science and defined behavior as the interaction of an autonomous agent with its environment, testing the behavior of ETs allows us to examine what Robert Brandon, back in Chapter 2, called "function in the ecological situation"—one of the six pieces of physical evidence needed for explaining adaptation.

Thus ETs, as primary representations, inform our evolutionary investigations by testing behavioral hypotheses of extinct or nonexistent animals. Behavior, as we've shown with our Evolvabots, is what selection "sees," the action in the game of life that we judge using the fitness function. However, because behavior doesn't fossilize, reconstructing it with ETs is a great way to remember the past.

REMEMBRANCE OF THINGS PAST

As I promised at the end of the last chapter, I'm not going to use Tadros, at first, to look at what we can learn by using ETs to study the biology of extinct organisms. I'm not going to look at fish nor am I even going to discuss backbones (at least not very much). Instead, our magical mystery tour of lost behaviors in the evolutionary landscape continues with the ET known as Robot Madeleine (Figure 7.4). Madeleine, scallop-shell shaped like the petit madeleine cake, is the first robotic creation named after a French pastry (Figure 7.5). Launched in 2004, Madeleine the robot served, just like Proust's tea-soaked madeleine the pastry, as the catalyst for explorations into things past,[15] lost vertebrates known as plesiosaurs who, with their

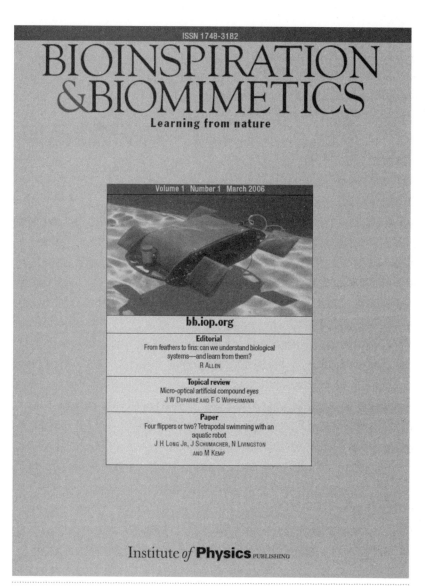

ISSN 1748-3182

BIOINSPIRATION &BIOMIMETICS
Learning from nature

Volume 1 Number 1 March 2006

bb.iop.org

Editorial
From feathers to fins: can we understand biological systems—and learn from them?
R ALLEN

Topical review
Micro-optical artificial compound eyes
J W DUPARRÉ AND F C WIPPERMANN

Paper
Four flippers or two? Tetrapodal swimming with an aquatic robot
J H LONG JR, J SCHUMACHER, N LIVINGSTON AND M KEMP

Institute *of* **Physics** PUBLISHING

FIGURE 7.4. *Robot Madeleine, a four-flippered Evolutionary Trekker.* Madeleine helped launch a new scientific journal, *Bioinspiration & Biomimetics,* in 2006. Madeleine is designed to represent aquatic tetrapods, descendents of the four-footed vertebrates that evolved on land and then evolved back to the water (see Figure 7.7). By varying the pattern of how she uses her flippers, we can test the hypothesis that four flippers, compared to two, produced swimming behavior with faster top speeds, quicker acceleration, and better braking. The cover image is used with permission of the Institute of Physics. I took the picture of Madeleine while she was going through her first shake-down cruise in the out-door pool of my friend John Keller.

FIGURE 7.5. *A petit madeleine, chocolate, left lateral view, showing its streamlined scallop shape as seen just prior to consumption.* Robot Madeleine is the first robot named after a French pastry. We recognize that petite madeleines don't have flippers and don't swim. However, the chocolate petit madeleines, in particular, are very tasty, like little moist cakes.

four propulsive flippers, were the giant top-level predators of the Jurassic seas over two hundred million years ago.

Well, perhaps not quite remembrance: we know of plesiosaurs only because they've left us their skeletons as fossils, not because we were around to see them when they still lived. The first plesiosaur was discovered by twenty-two-year-old Mary Anning in 1821 as she scoured the cliffs of Lyme Regis, a West Dorset coastal town on the English Channel. Anning's sea dragon was clearly a vertebrate—with its many vertebrae forming the great chain of bones along its axis— but was otherwise odd, with four large paddles instead of legs and a girdle of bones instead of gracile ribs amidships.[16] The Reverend Conybeare named it for science in 1824 as *Plesiosaurus*, from the Greek *plesio* (= close or near) and *saurus* (= lizard), and described it as a "comparison with the paddles of the sea turtle will exhibit such fresh analogies as to indicate that in respect of the various forms of animal extremities, the *Plesiosaurus* holds as it were a middle place between it and the *Ichthyosaurus*; for we may remark in the first carpal series of the turtle three bones not unlike those of the *Plesiosaurus*."[17]

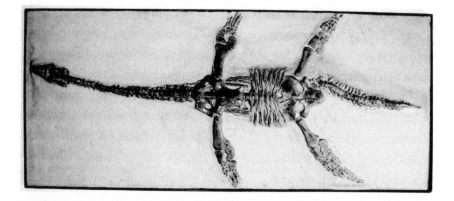

FIGURE 7.6. Plesiosaurus dolichodeirus, *cast of fossil.* Note the strangeness: a tiny head on a long neck, short body reinforced with robust bone, and, best of all, four large flippers of a form that appears to be shaped for doing the hydrodynamic work of an underwater wing. This cast, about two meters long, is from the Warthin Museum of Geology and Natural History at Vassar College. It was purchased in the nineteenth century by the college and was listed as Item 225 in the Wards Scientific catalogue of 1866. Photo by Rick Jones.

Richard Ellis, in his book *Sea Dragons*, notes that Conybeare's scientific description was followed by Dean Buckland's more famous 1836 construction: "To the head of a lizard, it united the teeth of a crocodile; a neck of enormous length, resembling the body of a serpent; a trunk and tail having the proportions of any ordinary quadruped, the ribs of a chameleon and the paddles of a whale."

Strangest of all is the fact that these descriptions are not flights of fancy. The darn things look pretty much as described in Figure 7.6.[18]

Although their utter strangeness makes plesiosaurs so compelling, in an evolutionary sense they were pedestrian. Plesiosaurs[19] are just one example of what Carl Zimmer has described as a repeated series of past and ongoing evolutionary experiments in which sea creatures descend from four-legged terrestrial vertebrates known as tetrapods.[20] From land to sea they go. Because of their heritage as terrestrial tetrapods, any vertebrate lineage that has crawled back to the sea, evolutionarily speaking, is called an aquatic tetrapod. Living aquatic tetrapods that you might recognize include whales and dolphins, sea turtles, penguins, otters, and seals and sea lions. And there are more!

Not even including amphibians, it's simply stunning how many times terrestrial tetrapods have spawned species that have returned to the sea and adapted to aquatic locomotion. Analyzing a beautifully preserved Late Cretaceous mosasaur (mosasaurs are yet another group of giant, extinct aquatic tetrapods), Johan Lindgren, a researcher at Lund University in Sweden, and his colleagues have shown how selection for enhanced swimming performance has apparently and repeatedly built streamlined bodies, vertebral columns reshaped to enhance and regionally control bending stiffness, and caudal fins with increased span.[21] Keep in mind that any pattern of convergent evolution (see Chapter 2) is excellent circumstantial evidence that a strong and steady selection pressure has been at work.

Convergent evolution creates one of those goose-bumps moments for biologists. I mean, how cool is it that dolphins and whales evolved from mammals to look like extinct ichthyosaurs and mosasaurs that were, in turn, independently evolved from reptiles? Not only does convergence provide great evidence for evolution by natural selection, but it also suggests that, in some kinds of situations, only a few options exist for pushing the performance envelope in the game of life.

When the game moves from land back to water, the aquatic tetrapods basically have two choices to overcome their locomotor roadblock: limbs or body axis. Frank Fish, professor of biology at West Chester University and an expert in the biomechanics of aquatic locomotion, has proposed a functional path for the evolution of aquatic tetrapods in mammals.[22] According to Fish, it's probably the case that swimming with appendages is the way that the shift back to water starts; this seems likely because almost every terrestrial mammal we see will swim using a variant of the "dog paddle" when they hit the water. Some lineages, such as seals and sea lions, stick with appendages, expanding and oscillating their rear flippers, in the case of seals, or flapping their front flippers, in the case of sea lions. Others—whales and manatees, for example—evolve their developmental programs to stop building appendages, losing the rear limbs and using the body axis, anchored by that pesky vertebral column again, to move the evolutionarily novel flukes up and down.

Pleiosaurs evolved along roughly the same track as seals, sea lions, and sea turtles, sticking with appendages as they readapted to life in the water. Here's a curious observation, one that makes me wonder and drives me mad with evolutionary might-have-beens: none of the living aquatic tetrapods ever use all four appendages to swim underwater—they use only two. The plesiosaurs, however, appear to have used all four limbs, which were modified into wing-shaped flippers (see again Figure 7.6). If four flippers were good enough for plesiosaurs to rule the seas as the top-level predators in the Mesozoic, why aren't they good enough now?

Let's ring the bell and bite the apple! We want to know: why, why, why? Why don't mammals and sea turtles alive today use all four flippers for propulsion? From a mechanical point of view, it sure seems like using four flippers for propulsion should be better in almost any way imaginable. If you think about each flipper as a propeller, then any agent—animal or robotic—using four flippers instead of two should be able to accelerate more rapidly, reach a faster cruising speed, and brake more quickly. So why wouldn't they? That, then, was the behavioral mystery within the context of evolutionary paths taken and not taken that we set out to solve with an Evolutionary Trekker.

BUILDING PROPULSIVE FLIPPERS

Here's where we need Robot Madeleine. We built her as a generalized aquatic tetrapod, with four identical flippers that propel her as she swims underwater. She's 0.78 meters long stem to stern and weighs twenty kilograms dry, roughly the length and mass of an adult green sea turtle or a small species of plesiosaur, minus the long neck (see Figure 7.6). Each of her flippers, called a Nektor by engineers, has the cross-sectional shape of a wing or, to be precise, a hydrofoil.[23] Each flipper is oscillated around an axis, "in pitch" as the engineers say, by separate motors inside Madeleine's hull. To avoid building an overly powerful super 'bot, we chose motors that would approximate the power density of vertebrate skeletal muscle, about ten watts per kilogram of body weight. And of course, Maddie has a body shaped like a petit madeleine pastry. From a biological point of view, this pastry shape would be described as bilaterally symmetrical, with a fusiform shape for streamlining.[24]

All of Maddie's features just mentioned were chosen with mechanistic accuracy in mind. So as convergent evolution seems to do for aquatic tetrapods, we focused on her locomotor behavior and the structures related to generating propulsion. To judge whether we'd done a good job of recreating an aquatic tetrapod, we rely on five of Webb's criteria for a model robot: biological relevance (criterion 1), behavioral match between the target and the model (criterion 2), mechanistic accuracy (criterion 3), level of structure (criterion 5), and substrate (criterion 7).

Perhaps the biggest complaint we get about Maddie is that she does not represent any species in particular, giving her low concreteness (or, conversely, high abstraction, Webb's fourth criterion). But . . . precisely! That's what we wanted, and Webb's criteria help us recognize and explain that Robot Madeleine can't—and shouldn't even be able to—do it all. In fact, that's why I named her after a French pastry. I didn't want to pretend that she was a robotic turtle, for example, as she has come to be named in the popular press. Can I have my pastry and eat it too? If Madeleine is not a robotic turtle, then how can I claim that she is a robotic plesiosaur? I don't. What I claim is that Maddie uses some of the same propulsive principles that we think both turtles and plesiosaurs use and used. Thus, Maddie's mechanistic accuracy is high for any aquatic tetrapod that flaps flippers to swim. Thinking about modeling as the process of representing, Maddie's behavior represents the behavior of turtles and plesiosaurs in the specific sense that she is about their size and swims with flippers.

Another important critique of Madeleine is that her flippers don't work in exactly the same way as the flippers of a turtle or a plesiosaur. This gets to the issue of accuracy of the model at the level of the propulsive elements. The motion of Maddie's flippers, or Nektors, as they are known, is unbiological, in no small part because their movements are much simpler than, say, the movements of the front flippers of sea lions.[25] Both Maddie's flippers and a sea lion's rotate in pitch during each stroke—but that's all Maddie's do. A sea lion's also roll as the front limb pivots about the joint of the shoulder. While the sea lion is rolling its flipper down and pitching it about its long axis, it is also yawing the flipper about the shoulder joint, moving the flipper's tip rearward toward its hip. To jump into the lingo of engineering, then, the flipper of a sea lion has three degrees of freedom, whereas Maddie's has only one. And even three degrees is still too simple! I've conveniently neglected to mention that the flippers change shape as they rotate, with joints at the elbow, wrist, and five fingers flexing and extending. Let's see, if we assume that elbow, wrist, and five finger joints are all simple planar joints, that adds seven degrees of freedom. In sum, each flipper has ten degrees of freedom from ten joints, and each joint has be actuated, controlled, and, as a group, coordinated.

Are you ready to give it a try? You can do it! But before you run off and build a better flipper, keep in mind that by jumping from one to ten degrees of freedom you'd be violating the KISS principle, at your own peril. You'd be doing the opposite of making the simplest device possible. Let's give that anti-KISS principle a name: Make It Complicated, Einstein, or MICE. With MICE, you'd need to create an internal skeleton with joints strong enough to withstand all of the hydrodynamic loads yet supple enough to bend without requiring too much force. You'd need to figure out how to move all ten of the joints that you've created: do you put motors out in the fingers and add bulk and weight, or do you run wires or hydraulic tubes from the inside of Maddie? Then you'd have to cover and embed the skeleton with a flexible, body-like material that maintains shape yet reconfigures as the flipper moves. And, once covered, that limb would need to allow you to get back inside for repair and maintenance.

Let's say that you took care of all of that mechanical engineering, using the MICE principle, and now you are looking at your handiwork. Now what? You've got to build another flipper, for the other side of . . . an underwater robot. Yikes! You've put the robot part on hold while you earned your PhD designing and building the biomimetic flipper. Okay, no big deal. Now you build a robot to anchor the two sea lion flippers and hold the batteries to power the motors. Done? Not quite. Now you need to design and test the software to control and coordinate the twenty motors. Twenty degrees of freedom aren't too much of a problem because you are just getting the flipper to flap down and back and then recover to the position needed to repeat the motion. Your software controller works great in the lab, in air, on the bench. But then you put the robot in the water. The moment it starts to flap and move, the long flippers undergo drag that tends to push back and bend the flippers. You realize that if the flippers are going to maintain a specific shape and change shape in a way that you specify during the stroke, you've got to have sensors at each joint.

Maybe this is why animals have proprioception, the internal positional sensory system that allows you to touch your nose with your index finger with your eyes closed. Thinking with your MICE hat on, you curse yourself for not making the flippers complicated enough from the get-go! You rip open the flippers and put potentiometers at each joint, and wire those twenty sensors to the computer controlling the motion of the flipper. You build a new software module that takes the sensor input as feedback to the motors. Now, when you tell the flipper to have a certain shape and position, you are certain it will do so.

Back at the pool, you put your system, now dubbed SeaLioTron, into the water and are overjoyed to see it swim slowly with a symmetrical and controlled motion of the flippers. Great. Now let's get SeaLioTron to turn. You'd thought about that aspect of control, but because you were working on the flippers, you saved that problem for later. When you revisit the videotapes that you've got of real sea lions, you notice that they have tremendous turning agility because they can bend, using their bodies as big rudders in the water. Because having a flexible body was not part of your project and raises all kinds of issues with the internal payload of batteries, motors, and computers, you either (a) call the SeaLioTron project a success and consider it done, (b) figure out how to turn SeaLioTron with flexible flippers and a rigid body, or (c) start a whole new project to build a flexible body-as-rudder.

Although I've made up this MICE counterexample, the idea that someone might tackle SeaLioTron is not complete fantasy. You would certainly encounter the problems that I've outlined and you'd probably, along the way, generate some really clever solutions that I can't fathom. And all that difficulty might be worth it: the SeaLioTron would easily produce a handful of PhD projects and, likely, a bunch of cool patents for flexible, actuated propulsors and multijoint neural controllers. You might even get NSF to fund the project. So what am I doing? I bring up SeaLioTron as an example of the MICE principle to make the following point about Maddie's one-degree-of-freedom Nektors: yes, Maddie's flippers are inaccurate as models of sea lion or sea turtle or plesiosaur flippers, *but* at the time that Maddie was built (2003 to 2004), they were the most bio-realistic flappers around (generating thrust by flapping with a flexible foil), and they are relatively easy to actuate and control. With the KISS principle, you do the simple stuff first. And the simple stuff turns out to be plenty complicated.

TWO FLIPPERS OR FOUR?

So back to the big question: why do all living aquatic tetrapods favor two powered flippers over four, when four seems to be the key to better performance? Even though ETs make answering the question easier than if we tried to explore the morphospace of flippers and the likely evolutionary processes, we still need to proceed with the extreme prejudice that a physically embodied robot will produce results that we failed to anticipate.

Let's first define our "simple," two-dimensional morphospace. The traits that we could vary in Madeleine were the number of flippers used and the pattern of flipper use. In the first set of experiments, however, we only changed and investigated the pattern of use in what we might call the neural-control space (Figure 7.7). Although that simplifies matters, the patterns are still awfully complex, a result of the fact that each of Maddie's flippers is independently controlled.

When one flipper reaches its most downward position, another flipper may be reaching its uppermost position. The difference in the time that two flippers take to reach the same position is called "phase." If two flippers are in phase, what we'll label as 0 degrees out of 360, then they are flapping together, perfectly synchronizing their swimming. If two flippers are out of phase, the easiest pattern to see is a phase of 180 degrees, like drumming a steady beat by alternating your left and right hands striking the table top. Unfortunately, there are many other ways to be out of phase, so then the combinatorial world of our experiment explodes. If you think about testing every ten degrees of phase, a crude resolution for this dimension that contains 360 degrees, that gives you thirty-six different conditions to go along with four flippers or—ouch!—more than a million total combinations (Figure 7.7, bottom). Cruel irony. Even with just two dimensions, number of flippers and the phase between them, we can't exhaustively explore this "simple" neural control space.

FIGURE 7.7. *Two flippers or four?* **(facing page)** Terrestrial tetrapods—mammals, reptiles, and birds—have repeatedly spawned lineages that returned to the sea. These aquatic tetrapods evolved in different ways, improving their submerged swimming performance over generational time by shifting from a terrestrial pattern of back-and-forth limb movement to an up-and-down or side-to-side motion. Those changes in motion are associated with a change from drag-based paddling to lift-based flapping. Living flappers, like sea lions of the genus *Zalophus* or the seals of the genus *Phoca*, use only two flippers for propulsion. In contrast, extinct flappers like short-necked plesiosaurs of the genus *Kronosaurus* or long-necked plesiosaurs of the genus *Plesiosaurus* have four nearly identical flippers that appear, from their wing-like shape and anatomical connections to the body, to have been used in lift-based propulsion. How swimming performance is connected to the motion of the flippers quickly becomes complicated, with millions of possible flipper patterns in four-flippered swimmers. This figure is inspired by the research of Frank Fish.

**Terrestrial
Tetrapods**

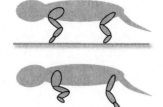

standing

bounding gait
using four limbs

swimming at surface
using four limbs
and a terrestrial gait

*Aquatic tetrapods have evolved multiple times
from various terrestrial ancestors.*

**Aquatic
Tetrapods**

submerged swimming using two limbs:
front flippers are propulsive

Zalophus

submerged swimming with four limbs:
all flippers are propulsive

Kronosaurus

submerged swimming using two limbs:
rear flippers are propulsive

Phoca

submerged swimming with four limbs:
all flippers are propulsive

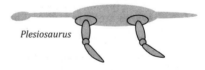

Plesiosaurus

**Neural Control
Space**

Dimensions, k, for the control of four-flippered swimming —

1. front limbs: left-right phase (0 to 360°, 10° increments, j = 36)
2. rear limbs: left-right phase
3. left limbs: front-rear phase
4. right limbs: front-rear phase

Number of possible combinations, n:

$$n = j^k = 36^4 = 1,679,616$$

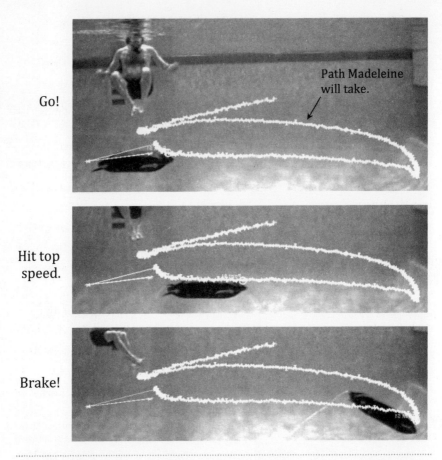

Go!

Hit top
speed.

Brake!

FIGURE 7.8. *Experiments with the Evolutionary Trekker, Madeleine.* To test the hypothesis that Robot Madeleine should swim faster and with greater acceleration using four flippers rather than two, we ran her through a series of experiments. In this example Madeleine is using her two rear flippers to start from a stop, swim as fast as she can, and then stop as quickly as she can. These images were taken from underwater video that had been analyzed to show Madeleine's position (point on her bow traced frame by frame to create the path) over the whole experiment. The snorkeler in the water (that's me, ahem) makes sure that Madeleine is stationary and located at a depth of two meters before the topside experimenters start the experiment.

In the face of this daunting complexity we sought the refuge of the KISS principle once again. In terms of number of flippers, we had three conditions to test: (1) two front flippers, (2) two rear flippers, and (3) all four flippers. Within each of these conditions, we varied the phase in the following ways. With two flippers, they flapped either in-phase (0 degrees) or 180 degrees out of phase. Simple. With four flippers, we borrowed a page from Frank Fish's evolutionary model and used four patterns of limb movements, called gaits, that are seen in terrestrial tetrapods: (1) pronk (all four in phase), (2) gallop (front in phase, rear in phase, front 180 degrees out of phase with rear), (3) trot (left front in phase with right rear, right front in phase with left rear, those diagonal pairings 180 degrees out of phase with each other), and (4) pace (left in phase, right in phase, left 180 degrees out of phase with the right).

We tested Madeleine using all eight gaits in the diving well at Vassar's pool.[26] Using an underwater video camera, we recorded Madeleine's movements as we accelerated her from rest, raced her to top speed, and then had her break as quickly as she could (Figure 7.8). While she was doing all of this, we measured her accelerations and energy consumption using an on-board three-axis accelerometer and electrical power monitor.[27]

The result? Two flippers are just as good as four in terms of top cruising speed (Figure 7.9). This was not what we'd expected. Another good theory dashed upon the rocky shore of empirical investigation! But this wreck is enlightening. Four flippers, in addition to not moving Madeleine along at any faster top speed, also takes twice the electrical power of two flippers. Two flippers good, four flippers bad—right? Not so fast. Four flippers are really handy if you want to accelerate quickly from rest, and using four gets you off the mark 1.4 times faster than using two flippers.

What we have here is called a performance trade-off. If cruising is your game, then you should use two flippers. If accelerating is important, use four. Trade-offs like this are the stuff of evolution, as the selection environment fiddles with what works best in a population at a

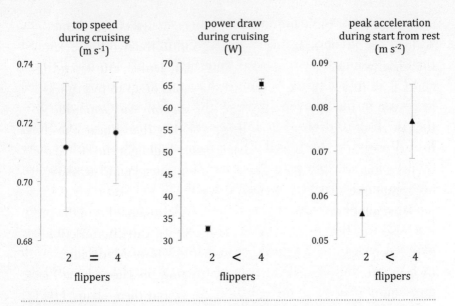

FIGURE 7.9. *Madeleine's behavior with two and four flippers.* To our surprise, Madeleine reaches the same top cruising speed using only two flippers. To maintain the same cruising speed with four flippers, Madeleine has to use twice the electrical power. Four flippers offer an advantage when Madeleine accelerates from a start. The points represent the means for three trials of each of four gaits. The error bars are the standard errors of the mean. Statistical tests back up all claims of a performance metric being the same or different.

given time and place. Change how schools of fish are distributed, for example, and maybe you have to cruise more to find your prey.

Let's get back to our motivating question: why do we see living aquatic tetrapods using only two flippers for propulsion? Based on the results from Robot Madeleine, my empirically educated guess is that living species tend to do a lot of cruising, relying most heavily on that aspect of their swimming behavior in the game of life. For example, green turtles, when they aren't asleep on the ocean floor, are cruising around, moving among beds of sea grass, which they visit for their vegetarian meals. Penguins cruise rapidly from shore to fishing grounds, where they dive and maneuver through schools of prey. Seals and sea lions, too, eat fish, although sea lions, in particular, are notable for their rapid turns and maneuvers, as we talked about with SeaLioTron. Sea lions may be an exception to the two-flipper-cruiser rule that we are sketching out. Frank Fish, Jenifer Hurle, and Dan Costa have measured centripetal accelerations of up to five times that of gravity in the California sea lion, *Zalophus californianus.*[28] Although those accelerations are angular rather than linear like those that we measured in Robot Madeleine, they highlight the fact that the sea lions don't need top cruising speeds but instead rapid accelerations to capture quick and elusive fish.

What about our extinct four-flippered plesiosaurs? From the physical evidence Robot Madeleine produced, we can circumscribe the likely scenarios. For small plesiosaurs the size of Madeleine, I can imagine them feeding as sit-and-wait ambush predators: hang in the water until something tasty swims close by and then—bam! Hit the accelerator and grab some lunch. Although this ambush behavior might work for smaller plesiosaurs, I don't think that it's possible for the giant short-necked pliosaurs like *Kronosaurus* or Predator X.[29] The trouble is that at ten or fifteen meters long, they are simply too massive to accelerate quickly. For the same reason that you never see a tractor-trailer beat a sports car off the line when the light turns green, you would never see *Kronosaurus* waiting and then lunging at a fish traveling by. Neither truck nor pliosaur can generate the mechanical power needed to launch their massive bodies quickly. When starting from a stop, their performance is constrained by (1) the amount of power that either internal combustion or skeletal muscle can produce and (2) their massiveness.

So what's a poor giant sea monster to do with four flippers? For the *Predator X* documentary that aired on the History Channel, I did some very crude calculations. The team of paleontologists that discovered Predator X, led by Dr. Jorn Hurum, estimates that Predator X was fifteen meters long. Using data that are available on the length and mass of great whales, I estimate that Predator X had a mass of about thirty-nine thousand kilograms, or thirty-nine metric tonnes. If Predator X accelerated at Madeleine's peak acceleration from rest, about 0.085 m s^{-2} (see Figure 7.9), it would move 8.5 centimeters in one second, a tiny fraction of its 1,500-centimeter total length. If you happened upon Predator X sitting still in the water, you'd have nothing to worry about, unless you swam right into its mouth!

If, however, you ran into Predator X while she was already moving, you might be in trouble. I'm guessing that Predator X cruised around and used its large flippers, like those on a humpback whale, to maneuver, redirecting its forward-cruising momentum into a feeding lunge of the type seen in blue whales, accelerating to one or two m s^{-2} to hit speeds perhaps as high as two to three m s^{-1}.[30] But what's Predator X predating? Unlike humpback or blue whales, which use their baleen plates to filter whole schools of small fish or krill out of sea water, Predator X, with its large teeth, was probably grabbing onto other large and relatively sluggish animals. The long-necked plesiosaurs may have been this short-necked plesiosaur's target.

Here's the best part: if Predator X cruised around looking for an unwary plesiosaur, then perhaps it only needed to use two of its flippers at any one time. Why waste the energy needed to flap all four if Robot Madeleine tells us that you won't swim any faster for the additional effort? Imagine if you could run a marathon on your legs *or* your arms. You could race until your legs were sore and then switch to your arms. This is straight out of the Department of Crazy Ideas. But based on what we now know about the physics of four-flippered swimming, switching between front and back propulsive systems makes sense.

For the truth police: we can never know for sure how plesiosaurs swam. Because they are extinct, their behavior is lost. No matter how accurate we make Robot Madeleine's limb anatomy, no matter how big or small we scale her, no matter how exhaustively we search the neural control space, the very best we can do using an ET is to talk about what is more or less possible. ETs help us circumscribe the plausible; what guides our judgment is the physical reality of behavior, that interaction of an embodied agent and its physical environment.

Acting as an ET, Robot Madeleine journeyed into the neural control space of four-flippered aquatic tetrapods to show us what swimming behavior looks like if you swim with two or four flippers. We compared benefits—speed and acceleration—and costs—power consumption. The differences in Madeleine's swimming behavior and energy use indicate that a trade-off could exist between high cruising speed and rapid acceleration. If you want both, you are going to pay for it in terms of the food you need to eat to make the energy that your behavior requires. Although our curiosity is not completely satisfied, we haven't gone mad—or have we?

BUILDING ROBOT MADELEINE

It certainly took a kind of collective madness to build a self-propelled biorobot like Madeleine—madness, induced by a shared vision and then a bunch of people with the know-how, the time, and the money to get the job done.[31]

Robot Madeleine was custom built for Vassar's Interdisciplinary Robotics Research Laboratory (IRRL) in about a year, from 2003 to 2004, by engineers at Nekton Research, LLC, in Durham, North Carolina. Working as Nekton's vice president of Science and Technology was Chuck Pell, cofounder with Steve Wainwright of the BioDesign Studio at Duke University in 1990. Under Chuck and Steve, BioDesign Studio had produced the early prototypes of bioinspired devices that would lead to innovations like Nektors and Transphibians, a class of unmanned underwater vehicles of which Madeleine was the first.

Having created a toy chest full of commercializable ideas, in 1994 Steve and Chuck teamed with businessmen Gordon Caudle and Jeff Bourne to form Nekton Technologies Inc., an independent start-up company that became Nekton Research, LLC, in 2000. Nekton, which acquired multiple government contracts under the guidance of president and CEO Rick Vosburgh, was acquired by iRobot Incorporated in

2008. As a result, Madeleine's four-flippered commercial descendants are now called iRobot Transphibians. Transphibians can be used for mine clearing, surveillance, and reconnaissance in physically challenging shallow and wave-swept marine environments.[32] You can see why Brett Hobson, Nekton's first ocean engineer and one of the patent holders, calls the Transphibian "Madeleine on steroids."[33]

Our shared vision for Madeleine began with Chuck sharing his vision. Before Nekton, during his time running the BioDesign Studio, Chuck drew a *Kronosaurus* on a napkin and started talking to me about building a life-size, swimming pliosaur. No big deal—a life-size and self-propelled pliosaur. As crazy as this sounds in the retelling, I didn't think that Chuck was nuts, at least not because of that idea. You have to understand that by 1992 Chuck had already demonstrated that he could use a Nektor to propel a surfboard. I first saw him do this at the Duke Marine Laboratory. He sat on the board, legs out straight on the board, and then cranked a large metal lever back and forth between his legs. Because you couldn't see the underwater Nektor attached to the shaft he was wiggling, his rapid exertions looked for all the world like he was trying furiously to tighten a bolt with a large socket wrench.

Always the enthusiastic, hands-on teacher, Chuck made all of us who were jeering him from shore give his swimming machine a try. Soon we were all hooked on flapping flexible foil propulsion, and we took turns wiggling the lever and zooming around the docks. The demonstration of Chuck's crazy contraption was exactly the kind of thing that got the business-minded among us fired up. Gordon started thinking about building a company that would make quiet, low-speed trolling motors for bass fishing. Steve saw a science toy that could be used in the sink or bathtub to demonstrate how fish make waves of body bends in order to swim. In fact, this science toy, which Chuck dubbed "The Twiddlefish," launched the toy company TwidCo, Inc., and Gordon and Jeff soon had little clown fish and sharks on the shelves of museum stores across the country.

But it wasn't until Nekton was created in 1994 that Nektors were put to work propelling craft. Brett, renowned in the ocean engineering community for his work on high-performance submersibles like Deep Flight, was Nekton's first employee. Arraying four Nektors around the circular belly of an ellipsoidal submersible, Brett and Eric Tytell built PilotFish, which quickly achieved the world record for underwater maneuverability, showing that Nektors had response rates twenty times faster than conventional thrusters, could provide thrust in any direction, and allowed PilotFish to use all available degrees of freedom.

By the time I approached Chuck about building a four-flippered robot in 2002, Nekton had left Nektors behind, according to Brett, leaving PilotFish on the shelf. Even though they were interested in exploring other body and flipper geometries, they didn't have the funding. So we got it from the National Science Foundation in an equipment grant that I wrote with my colleagues at Vassar, Ken Livingston, Tom Ellman, Luke Hunsberger, and Bradley Richards.

Money in hand, we began the design of Madeleine in earnest. Chuck already had Maddie's conceptual design drawn up when I came to work with the team at Nekton. Brett and Chuck were joined by engineers Robert Hughes, Ryan Moody, and physicist Mathieu Kemp. The Robot Madeleine project was important for the group because, said Brett, "[Madeleine] was the first piece of equipment that Nekton produced and delivered and somebody used."

What made Maddie useful as an ET was not just her glorious Nektors but also the way in which the Nektors were programmed to operate. This programming was Mathieu's masterstroke. Not only did he program the motor controllers, he also figured out how to have Maddie's on-board computer interface with all the sensors I wanted. This gave Maddie all of the capacities that Nick Livingston, Joe Schumacher, and I needed to conduct our flipper experiments at Vassar. In addition, Mathieu programmed Maddie to be autonomous and employ a two-layer subsumption hierarchy in 2004. This was the first time, to our knowledge, that an autonomous underwater robot had used Rodney Brooks's architecture (see Chapter 5). Although we never published Mathieu's autonomous design, proof of Maddie operating autonomously can be seen on the Australian Broadcasting Company's science and technology program, *Beyond Tomorrow*.[34]

In the same way that Chuck's swimming contraption was a great catalyst for what might come from Nektor-based propulsion, Maddie turned out be a great spokesperson for Nekton and what they could accomplish. "Video of Madeleine performing," reminisced Chuck, "electrified government sponsors who previously turned us away and who now said 'yes' to money for an even more powerful vehicle." For demos to the Navy brass, Brett and Mathieu would borrow Maddie, give her special fins, reprogram her, and turn her into an amphibious surf-zone explorer (Figure 7.10). As a Transphibian, "[Maddie] was targeted for high-energy environments," Chuck explained, "where bottom crawlers can deal with terrain and pelagic vehicles can't deal with surge. So this tactically important zone had no vehicle that could thrive in it." That's no longer the case, thanks to the crawling and swimming Transphibian.

OTHER EVOLUTIONARY TREKKERS

Robot Madeleine is not the world's first and only Evolutionary Trekker. Josh de Leeuw, a former student who has spent time, as an employee, working with and rebuilding Maddie, pointed out that in 1997 Tony Prescott, professor of cognitive neuroscience at Sheffield University, built a wheeled, autonomous robot to test ideas about the locomotor behavior of Cambrian invertebrates. With the simple intelligence to detect a track that it had made, Prescott's robot showed one possible neural mechanism that the ancient invertebrates may have used to create the foraging trails preserved as complex spiral patterns known as trace fossils.

To explore the neural control mechanisms involved in the evolution of terrestrial locomotion from aquatic vertebrates, Auke Ijspeert,

FIGURE 7.10. *Robot Madeleine as the first Transphibian, an amphibious surf-zone vehicle.* Maddie uses large flippers that rotate unidirectionally on land, like wheels, to move from the beach into the water. Once in the water, she switches to fin-flapping mode. Transphibians are commercially available from iRobot, Inc., which purchased Nekton Research, LLC, in 2008. Photo is courtesy of Brett Hobson.

associate professor and head of the BioRobotics Laboratory at École Polytechnique Fédérale de Lausanne, and his colleagues built a swimming and walking salamander robot in 2007.[35] The robot was programmed to walk and swim using neural chains of central pattern generators, which produce rhythmic activation of local body and limb motors. With a simple linear increase in the stimulation of the artificial nervous system, the robot transitioned smoothly from land to water, switching from standing to traveling body waves. Because the behavioral match and mechanistic accuracy of the robotic salamander are both high in terms of locomotion and nervous activation, respectively, this ET provides a plausible model for the evolution of the neural control of walking in vertebrates.

After all of our efforts to understand the evolution of early vertebrates and vertebrae, you won't be surprised to know that we have an ET project underway at Vassar to help us out. To explore the morphospace of just the number of vertebrae, we've built MARMT (Mobile

Autonomous Robot for Mechanical Testing), a swimming robot into which we can put propulsive tails with biomimetic vertebral columns. MARMT is another group project, with Jon Hirokawa, Sonia Roberts, Nicole Krenitsky, Carina Frias, Josh de Leeuw, and Marianne Porter all making important contributions to this Evolutionary Trekker. Rather than letting evolution create our variations, we simply vary the number of vertebrae from zero to eleven, keeping everything else the same, including the tail span. We then program MARMT to either swim steadily, using a variety of tail-beat frequencies, or to escape. MARMT accelerates faster and swims more rapidly with the stiffer tail that higher numbers of vertebrae produce.[36]

With Evolutionary Trekkers we learn how behavior varies in regions of morphospace (1) no longer occupied by living species, (2) never occupied by species, and/or (3) not visited by evolving robots. There you go, and there you are!

chapter 8

SO LONG, AND THANKS FOR ALL THE ROBOTIC FISH[1]

I'VE SEEN THE FUTURE, AND IT'S FULL OF FISH. NOT REAL FISH, of course. We've eaten most of those. I mean the robotic kind. Robotic fish are taking the place of the real ones. At the London Aquarium in 2005 you could have seen three bejeweled robotic fish, built by Professor Huosheng Hu and his team at Essex University, swimming on display. A swimming coelacanth, built by Mitsubishi Heavy Industries, made news in 2001 by making the rounds in the AquaTom in the Fukui Prefecture.[2] The French company RobotSwim has their new Jessiko autonomous robotic fish, ready for your swimming pool, local aquarium, or exploration and marine-monitoring mission.[3] Maurizio Porfiri, an engineering professor at the Polytechnic Institute at New York University, is building a robotic fish to herd real fish away from danger.[4] His NYU colleague, Professor Farshad Khorrami, has created a company, FarCo Technologies, that builds biomimetic robotic fish for industrial and defense-related situations.[5]

Why all the fuss about robotic fish? What's in it for you and me? Will a robotic fish become your best friend, save your life, or overthrow an evil dictator? Maybe. For certain, robotic fish will help us do what we can't naturally do: be underwater. As extensions of our hands and eyes, robotic fish are embodied-brain tools we primates will use to probe the aquatic depths. And as we've seen throughout this book, robotic fish are built with representations of bits and pieces of real fish, sometimes to learn more about fish, if you are a crazy biological cognitive scientist, or, most of the time, to build better machines, if you are an engineer with a job to do.

We all fall prey to the notion that evolution by natural selection is a better engineer than a hominid with a PhD from MIT. The implicit basis for this romantic view of nature as an engineer is that evolution perfects: "Evolution is a slow but sure process of perfecting design to give a life-form a natural advantage in a competitive environment."[6] Steve Vogel, in his book *Cats' Paws and Catapults*, explains the counterargument to this perfection bias beautifully: "Nature does what she does very well indeed. But—and here's the rub—why should she do so in the best possible way?"[7] Indeed.

Think about PreyRo. Even when that population of Evolvabots appears to have evolved a mechanically optimum tail stiffness, that doesn't mean that an average of 5.7 is the perfect solution forever and for everywhere. It's only the best solution—relative to others in the population—at that moment and place in the adaptive landscape. As we talked about in the last chapter, the adaptive landscape usually contains multiple peaks. In the face of such rugged terrain the best that evolution can hope to do is find the closest peak, the "local maximum" in the mountain lore of mathematicians. Even with evolution by natural selection in full hill-climbing mode, it can be pushed off course by random forces and prevented from getting underway by the historical constraints of the population's genetic history.

Instead, evolution suffices. It may provide just-good-enough solutions that aren't quite in time, but it doesn't even have to do that. (Nice work if you can get it, eh?) Selection, that judgeless judging environment with which an individual ceaselessly and unknowingly interacts,

plays a strong role in choosing the breeders who make the next generation. But by the time the next generation is on the scene, the world
may have changed, creating a new adaptive landscape that the previous, and different, selection environment could not anticipate. Nothing in the rule book for the game of life says that the playing field has
to be level or even has to stay the same. In fact, except in unusual places
such as the abyssal zone at the bottom of the ocean, the adaptive landscape for any population is better thought of as an "adaptive seascape,"
as suggested by Professor David Merrill in his eponymous book.

Given all of this complexity and contingency in the adaptive
seascape, do we really want to assume that we can look to the living
world to have solved all of our engineering problems and to have done
so perfectly?

No, we don't. But neither do we want to pretend that we have
nothing to learn from nature. Engineers understand this, and they
understand that they need more in their toolbox than just the dry
goods—stiff steel, flexible plastic, compressive concrete, and resilient
rubber—with which we've built the constructed world around us.
Phil Leduc, associate professor of mechanical engineering at Carnegie
Mellon University, works as a nanoengineer, manipulating proteins
within living cells, linking the mechanical behavior of one to the biological function of the other in order to design bioinspired nanofacto-
ries. Kenneth Breuer, professor of engineering and director of the
Fluid Dynamics Laboratory at Brown University, collaborates with
Sharon Swartz, professor of biology at Brown University, to study the
complex anatomy and behavior of flying bats and their extremely flexible wings as part of a larger project with engineers from the University of Michigan to build to a bat-inspired micro-air vehicle for the
US Air Force Research Laboratory. Melina Hale, associate professor
of organismal biology and anatomy at the University of Chicago and
a fish expert working on robotic fish, offers this insight: "with all the
tremendous work that has gone into designing and building robots,
we are still far from having one that functions as well as a fish."[8] It
holds for any bioinspired project.

INSPIRED BY FISH

What don't our robots do that fish do? Just about everything: fish have ways of sensing, navigating, and moving underwater, according to Maarja Kruusmaa, professor of biorobotics at Tallinn University of Technology in Estonia, that none of our robots have.

Maarja's team—an international one, comprising biologists and engineers from the Italian Institute of Technology, Riga Technical University in Latvia, the University of Verona in Italy, and the University of Bath in the United Kingdom—is designing a biomimetic lateral line and a robotic fish to take advantage of the new sensory capabilities the lateral line creates. We've discussed some of them: identifying the position of swimming companions in a school of fish at night, the location and proximity of an approaching predator, and the efficiency of self-generated water movement. All of these functions are beyond the scope of any single sensor currently made by humans. By building a biomimetic lateral line, Maarja and her team are betting that their robotic fish will be able to teach us land lubbers about physical structure in the aquatic world that we can't comprehend ourselves.

When I visited Maarja's biorobotics laboratory in spring 2009 I was struck by the ambition of the FILOSE (robotic FIsh LOcomotion and SEnsing) project.[9] The biomimetic lateral line will be placed on a self-propelled robotic fish (Figure 8.1) that controls its swimming performance by varying both neural control variables, like the frequency of its tail beat, and the mechanical properties of the body, such as stiffness.[10] The robotic fish will have to learn how the patterns of flow sensors distributed around its body relate to the external flow patterns and the detection of underwater objects; the FILOSE engineers have used a digital simulation to show that this is possible. This work is funded by the European Commission under their Seventh Framework Programme, which pursues the "European Union's Lisbon Strategy to become the 'most dynamic competitive knowledge-based economy in the world.'"[11]

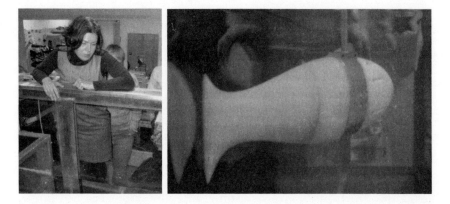

FIGURE 8.1. *The FILOSE fish robot at Tallinn University of Technology.* Professor of biorobotics Maarja Kruusmaa monitors the FILOSE fish robot in the flow tank, as seen on the right. This is an early developmental stage of the robot, driven by an external motor located above the tank, inputting power via the metal shaft. The large tail is an example of the KISS principle in action: it is a single piece of rubber acting as a robust actuator and rudder for the payload that sits in the rigid head. The biomimetic lateral line will be installed along the length of the FILOSE fish. The FILOSE project is multinational, with Professor Kruusmaa directing. The work is funded by the European Commission through their Seventh Framework Programme. Image on the left by John Long. Image on the right by Maarja Kruusmaa.

In the United States the National Science Foundation is funding work on fish to uncover novel ways to create highly maneuverable underwater vehicles. Malcolm McIver, professor of mechanical engineering at Northwestern University in Illinois, leads a team of engineers and biologists, including George Lauder of Harvard, studying the precision maneuvering of the black ghost knife fish.[12] Knife fish are so called because of their tapered and stiff body. Unlike other fish, knife fish don't bend their bodies to swim; instead, they ripple a long, thin fin that runs along their belly. This fin enables them to do something unusual among fishes: move vertically without having to point upward. By building a biomimetic robot, Malcolm's team has shown how this happens and that the function, once understood as a mechanical principle, can be used by engineers building underwater vehicles, even if those vehicles are otherwise unfish-like.

Using a process he calls "biologically derived design," James Tangorra, assistant professor of mechanical engineering and mechanics at Drexel University, builds robotic bluegill sunfish with George and Melina. Interested in efficient and novel propulsion, they focus on fins and their internal structures—thin, bifurcated rods called rays. In each of the sunfish's two pectoral fins, fourteen fin rays are independently controlled to bend, cup, and curl the fin. This level of structural control, which occurs dynamically during swimming, is unprecedented in engineering systems. By building fish-inspired robotic structures, the team has learned something new about real fish fins in the process: they bend using a novel structural mechanism involving the shearing of two parallel struts connected at one end.[13] With any flexible, animated structure, like a fin or a whole undulating body for that matter, neural control of its motion becomes the next frontier, as Melina explained when asked about design challenges facing her team and others building robotic fish: "we need to incorporate some sort of realistic understanding of sensation and sensory processing—taking in the sensory input and processing multiple inputs to, in a sense, decide on a motor output."

The FILOSE fish, the robotic knife fish, and the robotic sunfish fins projects show how fish inspire engineers. Engineers are also building robotic fish in China, France, Indonesia, Japan, Korea, Singapore, and the United Kingdom.[14] Among most of these fish-inspired projects we see three different approaches for getting started: (1) identify a *function* of a fish or function within a fish that is novel to engineering and can be reduced to a mechanical or mathematical principle, (2) identify a *behavior* of a fish or system of fish that is novel to engineering and can be reduced to an algorithm, or (3) identify a *structure* of a fish that is novel to engineering and can be duplicated in other materials. Clearly, novelties in aid of applications are what roboteers fish for.

For Frank Fish—who we met in the last chapter and who, funded by the US Office of Naval Research, heads a multi-institutional building of a robotic manta ray—nautical engineers' interest in fish has been constant and longstanding. "It all comes down to four words," he explained, "speed, maneuverability, efficiency, and stealth." But there's more to the story than just that: when I asked Frank why so many engineers were building robotic fish, without hesitation he said, "Because the Navy is funding them."[15]

BIOROBOTIC FISH PREPARE FOR WAR

The US Navy, through the Office of Naval Research (ONR), has been funding work on fish and dolphins since the office's inception in 1946, leveraging academic research to help engineers understand aquatic propulsion.[16] Having had two ONR research grants in the 1990s with my mathematics collaborator from Lafayette College, Rob Root, I've benefited from the Navy's interest in fish and robotic fish.

Rob and I conducted experiments and built mathematical models of swimming fish with much help from students like Craig Blanchette, Nick Boetticher, Hayden-William Courtland, Vynette Haultaufderhyde, Wyatt Korff, Nicole Lamb, Nicole Librizzi, Matt McHenry, Karen Nipper, David Paul, William Shepherd, Eamon Twohig, and Stephanie Varga. We were able, with additional help from our collaborators Peter Czuwala, Lena Koob-Emunds, Tom Koob, and Chuck Pell, to develop theory about the importance of body stiffness in swimming fish.

The punch line: yes, you can tune a fish. Fish tune their bodies by changing stiffness. Because stiffness is proportional to the speed[17] at which a wave travels or vibrates, using muscles to stiffen the body will drive the fish's flexural waves faster. When you make undulatory body waves, you perform the mechanical work of transferring the body's momentum to the water, and when the body's flexural waves move faster, you are generating more power.

When we had some evidence that real sunfish were using their muscles to alter body stiffness, ONR gave us a chance to build our first self-propelled robotic fish.[18] Matt McHenry and I were anxious to make Nektors in the image of our study species, pumpkinseed sunfish; Chuck Pell, one of the Nektor's co-inventors, kindly agreed to share his invention and his talents. He created a mold using a dead sunfish, from which he fabricated five identically shaped sunfish-Nektors. Each sunfish model was made out of a slightly different formulation of PVC rubber, which is how Chuck varied the models' material stiffnesses over a range that included the stiffness values we had measured in real sunfish.

With help from Vassar Professor Bob Suter, Matt and I developed a way to swim the sunfish model in a flow tank—kind of a treadmill for fish. For a given and constant frequency, each sunfish model swam upstream in the initially still flow tank. We increased the flow speed until the sunfish model could just swim in place, moving neither upstream nor downstream, balancing the drag and thrust. After doing this for all of the models and a bunch of frequencies, we found that stiffer sunfish swam faster.[19]

Recently the Navy's biorobotic research efforts have been joined by those of another federal agency, DARPA, the Defense Advanced Research Projects Agency.[20] In addition to an ONR-like goal to improve naval technology, DARPA's role in national defense ranges more broadly, across the services, as shown by their mission statement: "DARPA's mission is to maintain the technological superiority of the U.S. military and prevent technological surprise from harming our national security by sponsoring revolutionary, high-payoff research bridging the gap between fundamental discoveries and their military use."[21]

DARPA funds numerous robotics projects that use animals for inspiration. One such DARPA-funded project was led by Robert Full, professor of integrative biology and director of the Poly-Pedal Laboratory at the University of California at Berkeley. Bob, an expert in biomechanics and animal locomotion whom you may remember from Chapter 6, teamed with engineers at Boston Dynamics, Inc., the Illinois Institute of Technology, and the Robotics Institute at Carnegie Mellon University to build a six-legged robot inspired by insects. Called RiSE (Robots in Scansorial Environments), the 3.8-kilogram robot successfully scaled a three-story vertical concrete wall, untethered.[22]

RiSE is one of many successful robotics projects that DARPA has funded. In an effort to propel the development of autonomous battlefield vehicles, they created the DARPA Challenge, open to any team. In 2004 a robotic car, Stanley, the creation of Stanford and Volkswagen engineers, used its superior path planning, object detection/avoidance, and navigational skills to run the fastest time in a 138-mile course across the desert. Three years later, in the DARPA Urban Challenge, Carnegie Mellon University's autonomous car, Boss, won the $2 million prize when it was the fastest robot to navigate a cityscape, filled with real pedestrians and human-driven cars, without violating the California driving laws. Although these well-publicized challenges featured automobiles, DARPA continues to fund research into animals and bio-inspiration.[23] Last time I checked, it looked like you could make pitches to DARPA for bio-inspired and robotics research to the officers running the following programs: "Biologically-driven Navigation (solicitation number DARPA-SN-11-07)," "Deep Sea Operations (DARPA-BAA-11-24)," and "All Source Positioning and Navigation (DARPA-BAA-11-14)."[24]

With all this potential federal funding for defense-related, bio-inspired robots in the United States, I guess it shouldn't come as a surprise that other countries have robotic warfare programs. But I was flabbergasted, I admit, to learn just how many. According to robotics engineer Ron Arkin, professor of engineering and renowned roboticist at Georgia Technological Institute, fifty-six nations are developing robotic weapons.[25] And that forces an uncomfortable reality check: even though Arkin doesn't discuss underwater robots in his book on robotic warfare, I too am working—at least indirectly—for the military.[26]

PLAYING WAR

Growing up, I drove my mom crazy because not only did I play war incessantly with my buddies, Fritz and Karl von Valtier, but I also read about war. I read everything I could get my hands on, and I remember how impressed with myself I was when I had been through every book on war in the library of North Hills Elementary School. This was the 1970s, the middle of the Cold War with the Soviet Union, and after exhaustive scholarship, eleven-year-old John decided, to my mother's horror, that nuclear weapons had taken all the fun out of war. The Cold War, with tanks and troops lined up just looking at each other, was boring—nothing happened. What would General Patton say, I asked? I just couldn't see how we could be "at war" without any action.

My mother, who had marched against the Vietnam War, told me that if I got drafted, she'd shoot me in the foot. I'm not kidding. She said so repeatedly and passionately. During what my family called my "Spock phase," I fought her passion with my deadpan version of Vulcan logic. I explained, without emotion, of course, that I wasn't interested in going to war to die but rather to fight the bad guys and protect our country. When someone attacks your ship or your planet—I mean country—it was only logical to defend yourself. For some reason this line of argument failed to penetrate. I guess she was Bones to my Spock: illogical.

World War II, I told her, was my favorite war. She nearly died when I said that. How could her son—or any one for that matter—have a "favorite" war? "Well, mom," I remember explaining, "World War II had action, on multiple fronts, with different enemies who used different strategies and tactics." It was modern, with incredible technological developments in rocketry, aviation, and naval engineering. Best of all were the aircraft carriers and the strategic battles of the Pacific Theater! I bought books on carriers and carrier warfare. When I sat at the dinner table and explained to my older sister, Ann, that she needed to know that our pilots loved the Grumman F4F Wildcats, even though they were too slow to duel the Mitsubishi Zeroes head to head, because they had self-sealing fuel tanks and armored cockpits, I think that my mom got what I was really into: technology. Boys and their toys. She backed off—and didn't mention shooting me in the foot again.

I forgot about World War II, and Spock, for good measure, thanks in no small part to junior high school and the rage of the hormones. Studying primate social behavior and mammalian reproductive biology became the pursuit of the neighborhood boys. On the latter subject, we found our school library to be woefully inadequate. Fortunately, both of Fritz and Karl's parents were physicians, and the Drs. von Valtier kept their medical library at home well stocked and unlocked. We read widely and voraciously, appreciating the fact that anatomical texts were illustrated, and we thought it best not to trouble the good doctors with the knowledge of our library visits, even though I'm sure they would've been proud of the fact that we were inspired and independent scholars.

I was proud, for my part, of learning to drive. I impressed my instructor, and myself, when I was the only student in drivers' education who could handle a manual transmission. My instructor, however, was already married, so I needed a different means of drawing the attention of the eligible girls in my class. Where facility with a stick shift failed, the opportunity provided by a license and the family car worked. I turned sixteen, got the keys, and fell in love.

A year later I was still in love, dating the same girl, and was now the owner of a 1169 cc, four-speed, three-door hatchback '74 Honda Civic that my Uncle Pete had kindly saved for me in his garage. I also had a job, washing dishes at Mateo's Pub, to support my bliss. With such happiness, imagine my surprise when, at age seventeen, I started thinking about war again, and this time for real. I wanted to be an officer in the Navy. It was simple. I wanted to go to the Naval Academy at Annapolis, get a four-year college degree there, and then, I don't know, go fly Grumman F-14 Tomcats on the aircraft carrier USS CVA(N)-65 *Enterprise*.

Mom went, as they say in the military, ballistic. Still unaware of the irony, she again discussed the confluence of ordnance and my distal appendage as a deterrent to my service in the armed forces. But, I countered, this wasn't the draft, the press gangs that she decried in years of old. That, and the Vietnam War, were long over. I was choosing to go into the Navy to pursue my education (and maybe to mess around in boats too). Then I'd have a degree, a steady job, and after twenty years as an officer I could retire and start my own marine biology business, like Jacques Cousteau. A man, a plan, a canal: Panama!

My working-man argument failed to gain purchase with Mom. Let's sum her position up: war is bad because I would be trained to kill. I needed to switch tactics. Okay. Hmmm. Too bad I couldn't find a way to be trained to save instead. Wait! The Coast Guard, which I immediately dubbed, for marketing purposes, the "humanitarian branch of the military." Mission: search and rescue, drug interdiction, aids to navigation, and, my favorite, iceberg patrol.[27] Mom was on board.

The Coast Guard maneuver gave me some sea room, so to speak. I was allowed to take the admissions test for the Navy's Nuclear Power School (passed) and applied for and won an NROTC (Navy Reserve Officer Training Corps) scholarship to the School of Engineering at the University of Michigan. Mission accomplished? Not quite. It turns out that as I learned more about the Coast Guard, and the Coast Guard Academy in particular, I decided I didn't want Navy—I wanted Coast Guard. Only one problem: I didn't get into the academy.

I was crushed. My boyhood dreams and now teenage ambitions were dashed upon the rocks of reality. Tormented by the Sirens, I resigned myself to become a naval officer through Michigan's ROTC program. Mom would just have to deal with the Navy.

But she didn't have to. About four weeks after I'd received the Academy's letter stating that I was not a principal appointee, I received another saying that I had been named as an alternate, and a position had become available. Would I still be interested, they asked, in joining the "hard core about which the Navy forms?" By gum, an officer and a gentleman, I would be!

After high school graduation I sailed off, with Mom at the helm in her Chrysler K-car land cruiser, to the Coast Guard Academy in New London, Connecticut. Next day, my hair was shorn, I bravely kissed my mother good-bye, and then I joined the military. But not for long.

After the six-week boot camp, I found a way to get out, honorably. And I was glad to do so. But not for the reasons you might think.

I didn't mind the endless push-ups. I loved the seamanship training and messing around on boats and ships. I hated having to learn how to waltz. (I'm not kidding. Remember: an officer and a *gentleman*.) The food was okay, but I wasn't allowed to look at it (I'm not kidding about that, either). Even being hazed as a lowly "swabby" by the upperclassmen was tolerable. For example, by proclamation in our cadet handbook, we were commanded to know the correct time at all times. But it also stated that the correct time was impossible to know. With this paradox in mind, I was trained to spit out, rapid fire, the one and only proper response to the simple question, What time is it, mister? "Sir, I'm greatly embarrassed and deeply humiliated that due to unforeseen circumstances over which I have no control, the inner workings and hidden mechanisms of my chronometer are in such great inaccord with the sidereal motions by which time is generally reckoned that I cannot with any degree of accuracy state the correct time, sir. Sir, I can state without fear of being too far in error that the approximate time is [glance at watch and state the time in military time]."

With all of this fun playing war, what would cause me to abandon ship? Ronald Reagan. I entered the Academy in 1982 as one of 232 cadets, sir. What I learned, after swearing in, was that my class, the class of 1986, was smaller, by exactly 100 cadets, than the class of 1985. The Coast Guard, for historical and strategic reasons under the Department of Transportation rather than the Department of Defense (see: humanitarian branch of the military!), was subject to drastic budget cuts under Reaganomics. A survey was conducted amongst the swabbies regarding preference for majors. Not interested in marine engineering, ocean engineering, mechanical engineering, electrical engineering, physics, or prelaw for any of my three choices, I thrice checked, with a regulation number-two pencil, the small square boxes of my destiny: marine science. The next day it was announced: the Academy would be cutting the major in marine science. Hit! You sunk my battleship!

STUDY WAR NO MORE

Whereas my inner child wished for exciting wars and my young adult pined for a seafaring career, my middle-aged adult, looking back, is mortified with both desires. How easy it was to romanticize every aspect of war when you aren't in it. It wasn't until I transferred to the College of the Atlantic (COA), where I earned my undergraduate degree, that friends and mentors helped me challenge my implicit acceptance of the Roman poet Horace's call to battle: *Dulce et decorum est pro patria mori* ("To die for the fatherland is a sweet thing and becoming"[28]).

One mentor was Dr. Ted Grand, an expert in the functional anatomy of mammals at the Division of Zoological Research at the National Zoological Park (NZP) in Washington, DC. His colleague and my COA adviser, Sentiel, a.k.a. "Butch," Rommel, had persuaded him to take me on for an internship. Butch, a former naval officer and expert in navigation of vessels and people, had course corrected my life by helping me reckon, after leaving the Coast Guard, the heading back to the ocean. At the water's edge on Mount Desert Island, Maine, Butch had me studying the biomechanics of marine vertebrates. Apprenticing at the NZP, reasoned Butch, was staying the biological course because Ted would help me improve my skill as a functional anatomist and my ability to ask and answer evolutionary questions with a quantitative approach.

In the beginning what neither Butch nor I appreciated was that each time Ted led me through the dissection of an animal, he was also guiding me on multiple levels, challenging me to make my implicit assumptions explicit and justify my positions. I don't recall if it was over a wildebeest or a hippo, but he quickly learned of my interest in war and my background in the Coast Guard. I'm sure he must have understood my romantic/heroic naïveté, because he quickly took me to several used book stores to expand my military library. I soon learned that my elementary school's reading list hadn't included the likes of Wilfred Owen, the British soldier and poet who wrote from the World War I trenches of a fellow infantryman gassed by a German shell:

> In all my dreams, before my helpless sight,
> He plunges at me, guttering, choking, drowning.
>
> If in some smothering dreams, you too could pace
> Behind the wagon that we flung him in,
> And watch the white eyes writhing in his face,
> His hanging face, like a devil's sick of sin;
> If you could hear, at every jolt, the blood
> Come gargling from the froth-corrupted lungs,
> Obscene as cancer, bitter as the cud
> Of vile, incurable sores on innocent tongues—
> My friend, you would not tell with such high zest
> The old Lie: *Dulce et decorum est*
> *Pro patria mori* [29]

I hadn't read Michael Herr's "Dispatches," a first-person account of Vietnam from what we would call, today, an embedded reporter:

Whenever I heard something outside of our clenched little circle I'd practically flip, hoping to God that I wasn't the only one who noticed it. A couple of rounds fired off in the dark a kilometer away and the Elephant would be there kneeling on my chest, sending me down into my boots for a breath. Once I thought I saw a light moving in the jungle and I caught myself just under a whisper saying, "I'm not ready for this, I'm not ready for this." [30]

These days, Ted points out that although I have left behind the facile romanticism of war as the hero's journey, what really appears to have me bothered, he says, is the secretiveness that is essential to waging war. He might be right. I hate telling and keeping secrets, as my children will tell you, because someone always gets hurt when you withhold information. That parental stance informs my professional life too.

In the summer of 1999 Rob and I, at ONR's request, went to the Unmanned Untethered Submersible Technology (UUST) conference hosted biannually by the Autonomous Undersea Systems Institute. With us was Peter Czuwala, the engineer working in my lab, who presented on analytical models of swimming fish that he and our students, Craig Blanchette and Stephanie Varga, had spearheaded. Any moral concerns that our students voiced about working for the Navy I addressed by explaining that all of our work was available in the public domain.[31] I was proud, I said, to have the Department of Defense supporting nonproprietary work on fish. No secrets.[32]

For his part, Rob's line in the sand was making weapons. Toward the end of this UUST meeting, our program officer, the person running the ONR's Bioengineering Program and overseeing our research grant, sat all of her grantees down in a room. We had heard rumors of the pressure that she and other ONR administrators were getting from the admiralty to justify all of this academic, basic research. "In future research proposals to ONR," she said, "we want to see explicit reference to how your work will help us make better weapons-delivery platforms." Rob and I looked at each other and said, "We're done."

EVOLVING ROBOTS FOR THE MILITARY: THE NEW ARMS RACE IS NO SECRET

But we're not done. As long as we work on fish and robotic fish, even if we publish openly, we are part of a new arms race, a race among fifty-six countries working to weaponize robots.[33] Given that everyone is building robots for war, we'd be wise to heed DARPA's mission: "prevent technological surprise from harming our national security." One way to avoid surprise is to consider the obvious. By "obvious" I mean all that is available to you and me, those of us without special clearances, the not-secret information that, when you look at it from a different perspective, may allow you to guess about what is happening in secret. We'll take a look from our new perspective of evolving robots.

Although most of the military robotic systems I know about appear to be remotely controlled, with a human in the control loop, some are semiautonomous. It won't be long before fully autonomous robots are in operation because they can operate faster and more accurately than humans.[34] Their improved performance on the battlefield will drive innovation in that direction. Speed kills.

After that the logical direction, as I see it, is to move toward robots adapting their behavior as the battle wages. Behavioral adaptation is what makes rag-tag rebels so hard to beat in a protracted war. As part of the rebel alliance, you may be outmanned and outgunned, but every enemy has a weakness; if you can figure it out and take advantage, then you have a chance. For example, the US military is vulnerable to attack from improvised explosive devices, simple but deadly weapons that disrupt vehicular patrols in Iraq and fuel delivery in Afghanistan. The fastest way to adapt is through learning, and if you are a robot this means getting feedback about your performance that changes your onboard software. Behavioral adaptation is already well established in robots, with multiple methods for learning. One such adaptive learning algorithm is called an "adaptive neural control chaos circuit," invented by Poramate Manoonpong, professor of physics at Gottingen University in Germany, and his colleagues for rapid and reversible learning in changing environmental conditions.[35]

But learning or evolving software only takes you so far. The next step will be to have adaptation of the body, the hardware itself, by setting up selection to act on a population of Evolvabots. "Hod Lipson is the state-of-the-art for mechanical adaptation," explained former student and now colleague Josh de Leeuw. I agree. Hod, associate professor of many things (engineering, computer science, robotics) at Cornell University, takes a bio-inspired approach in order to engineer machines that can build other machines. Hod's lab has designed and built embodied robots automatically from digitally evolved scenarios and, along with his colleague Josh Bongard, assistant professor in computer science at the University of Vermont, created robots that sense self-injury and respond by altering their body and behavior.[36] Bongard takes an approach that he calls "artificial ontogeny," allowing a robot to combine learning over its lifetime while it is embedded as an individual in a population of evolving robots.[37]

At first, the idea of machine-making machines—replicating themselves, reproducing novel offspring, or making offspring for someone else—may seem far-fetched. But as Adam Lammert, one of the inventors of Tadro (see Chapter 3), pointed out to me recently, self-replicating machines were shown to be feasible as early as 1957.[38] Lionel and Roger Penrose, at the University College, London, demonstrated that recognition systems and subunits needed for replication could be built into the body of a physically embodied model organism. These models, made of plywood, were "creatures" with levers that prevented joining other creatures unless the proper mechanical signature was present. Once conjoined, two creatures, or a creature and subunits, could undergo fission to create two replicants.[39] Today Hod uses rapid-prototypers, machines that create three-dimensional parts of nearly any size and shape from software instructions.

Thus, the final step in making military robots will be to have them make their own robotic offspring as part of evolution on the battlefield. Robots will evolve their brains and bodies in response to the dynamic chaos of the moment in order to carry out their longer-term mission. Evolutionary adaptations will occur in a population of military Evolvabots with feedback from the battlefield environment using a fitness function the battlefield commanders generate. That fitness function might reward performance such as rate of target detection, rate of target hit, probability of survival, and robustness, or the ability to continue to operate once damaged.

When I presented this fitness idea to Chuck Pell, he disagreed: "The basic principle for robots in war should be this: [robots should be] unmanned, expendable, and cause maximum damage." When I pointed out that a simple fitness function of maximum damage doesn't leave room for evolving very complicated robots, he said that

FIGURE 8.2. *Pell's Principle in action.* Swarms of simple, expendable robots over-whelm more complicated systems. In any environment swarms succeed by getting close and manipulating sensors, motors, and communications. The robotic systems represented here are based on real systems under development. Note the bio-inspired and diverse designs. Watercolor pencil sketch by Charles Pell.

was the point. Chuck's perspective is that complicated robots are expensive, and that's a problem. With expensive robots, he explained, commanders won't want to lose them, and those in charge will alter their tactics accordingly. The inescapable consequence, which I'll call Pell's Principle, "is that robot capabilities are proportional to cost, and cost is inversely proportional to expendability." The logical conclusion of Pell's Principle is to build and use swarms of simple, small, expendable robots (Figure 8.2).

"You can't stop all the little robots," said Chuck, "robots like Micro-Hunter."[40] MicroHunter was a microAUV (Autonomous Underwater Vehicle) built using the same cycloptic helical klinotaxis system that we used to build Tadro (Figure 8.3; MicroHunter was introduced in Chapter 4). But whereas Tadro is a surface swimmer, MicroHunter swims underwater, spiraling its way to a light source from anywhere you could put it in an Olympic-sized pool. MicroHunter, funded by a DARPA contract to Chuck as an employee of Nekton Research (see Chapter 7), and Hugh Crenshaw, then professor at Duke University, caught their program officer's eye when the duo reported that it had a 100 percent success rate finding the target.[41] "Nothing is ever that good," said Chuck, "so they sent statisticians and a former Navy SEAL to investigate."

SEALs, the US Navy's special operations experts, are world renowned for their abilities underwater.[42] So even though Chuck and Hugh knew that MicroHunter could hit the three-meter light target when unchallenged, they figured that a group of four MicroHunters, swimming slowly, wouldn't stand a chance with a special ops SCUBA diver in the pool. To everyone's surprise and the SEAL's chagrin, the MicroHunters

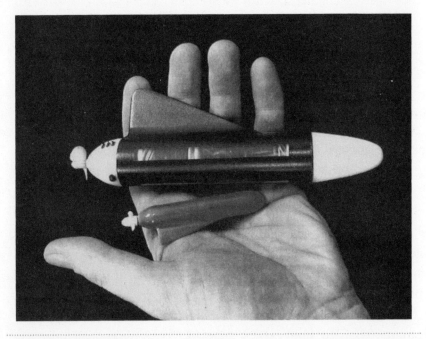

FIGURE 8.3. *MicroHunter, a fully autonomous micro-aquatic robot with just one moving part, a propeller.* MicroHunters, seven- and seventy-gram versions shown here, were developed by researchers and engineers at Nekton Research and Duke University with funding from DARPA's MicroSystems Technology Office, Distributed Robotics Effort. Photo by Charles Pell.

and the SEAL played to a draw, with the MicroHunters hitting the target 50 percent of the time in six three-minute trials. That's seriously excellent performance for a piece of embodied intelligence with but a single sensor and a single degree of freedom on the motor output side. Now try to imagine defending against not just four MicroHunters but a swarm of fifty. Your only defense may be evolution.

We've had enough theory and practice of evolution in action in this book that I'm guessing you'll be able to guess what I'm about to say. Here we go: any military that evolves robots on the battlefield will likely do so using the following principles:

1. *Robots are expendable.* Given Pell's Principle (see page 223), the only way to get large numbers of robots on the battlefield is to have them be expendable, and that usually means inexpensive. Large numbers also ensure that sufficient variation is in place to allow selection to act on the population. Large numbers also serve the tactical advantage just explained with regards to overwhelming the adversary with a swarm of simple autonomous agents.

2. *Robots are simple.* This, too, follows from Pell's Principle. The way to make robots expendable is to make them simple. Simple also usually means inexpensive to make and quick to produce. Employ the KISS principle in your design. Find the minimum brain, body, and behavior needed to seed your population. Choose which characters evolve.

3. *Robots evolve quickly.* Given that generation time is one factor that limits the rate of evolution, make the generation time short. Short generation time will minimize the response time between a change on the battlefield and the change in the behavior and hardware of the population of robots.

4. *Robots evolve in small cohorts (small and genetically isolated subpopulations).* Given that evolutionary change occurs rapidly in small, isolated populations, create many small companies of robots rather than a single large battalion. Keep in mind that random influences will dominate if the population is too small. Note that this may, at first, seem to run counter to principle 1 and using large numbers. You can have multiple populations or companies in simultaneous operation.

5. *Robots diversify in generational time.* Given that evolutionary change will be rapid with principles (3) and (4), then allow the companies of robots to speciate, to evolve along different evolutionary trajectories. Diversification will allow more and different kinds of adaptation to occur simultaneously, thus increasing the chances of both tracking changes in the environment and finding the best solution at a given time and place.

That's the offensive view of military Evolvabots. What about defense? How would you stop an army, navy, or air force of evolving robots? Keep in mind that even if it's your militia of Evolvabots, you will need ways to constrain and control them too. This is starting to sound like nearly every story and movie on robots in which they rise up and break the shackles of their creators to take over the world. Although we aren't making a movie, let's run with the plot line anyway. To avoid military defeat or robotic overthrow, here's what you do:

1. *Limit initial production prior to battle.* Control numbers and types of robots. Constrain raw materials. Limit, reduce, or eliminate energy sources. If the population is yours, you may want to design in hard production or run-time limits to avoid the enemy co-opting the group.
2. *Limit reproduction on the battlefield.* Because reproduction is key to the evolutionary process and usually involves some vulnerable moments, like finding mates and creating offspring, you should target these situations. Also, target the machine-making machines, because they need to be on the battlefield to keep generation times short. Limits to production (see item #1, above) can also be employed on the battlefield by cutting supply lines.
3. *Limit repair on the battlefield.* Injury provides another vulnerable situation. If robots are self-repairing, their function will be impaired. Capture low-performance agents preferentially. If the injured robots are being repaired by other agents, target the repair teams.

4. *Evolve predatory robots.* If you or the enemy employ Pell's Principle, you'll need to be prepared to capture or destroy swarms. For starters, you'll need to let your evolving predators have the capacity and capability of filter feeders like baleen whales. Consider behavioral adaptation first in your predators because the shorter generation time of the prey will limit opportunities for hardware evolution in the predators.

5. *Make complicated robots.* That's right. Want to control your own robots? Make them complicated. Because complicated usually means expensive, you are likely to have only a few of them. You will also hesitate to send them into harm's way, as Chuck was predicting. Furthermore, complicated robots will never take over because the laws of probability virtually guarantee their failure. If every component has, let's say, a 99 percent chance of not failing on a given mission, that sounds pretty good, right? But what if you have two such components in your robot? By the law of independent probabilities, we take the product of the two: 0.99 x 0.99 = 0.98. Not bad. A 98 percent chance of the system, composed of those two components, not failing. Now give your system two thousand components, not unreasonable for some of the more sophisticated fish robots I've seen. That's 0.99^{2000} = 0.0000000002. You've got no chance—your robot will fail! The way we keep complicated machines like airliners and space shuttles in business is by building components with lower failure rates (0.99999), engineering redundancy into the machine's critical systems, and inspecting and replacing parts before they fail. Bottom line: to ensure control of your Evolvabots, make them out of many crappy components.

COMMAND AND CONTROL

My Hollywood alarmism about evolving robots on the battlefield may have you thinking that all of this warfare stuff is just fantasy nonsense. Maybe it is. But let's pretend that you are an admiral and you have to make that call: are military Evolvabots nonsense or good sense? Will some other military surprise us with Evolvabots in battle? This matters because you, as admiral, need to make practical decisions that have long-term consequences. Do you put your limited resources into an offensive Evolvabot development project? Or do you put resources into Evolvabot defensive countermeasures? Keep in mind that if you do use resources on Evolvabots, you have to cut the budget of other tactical systems. How can you be sure that Evolvabots will ever be a serious risk and worth the cost of development or countermeasures?

You can take DARPA's tack and examine feasibility. Presented with the idea of commanding a fleet of evolving robotic fish, for example, you might want to assess one of the most important aspects of any battle: communication. If no one can figure out how to communicate with a swarm of underwater robots and adjust plans as the battle commences, then you probably don't have much to worry about.

Communication before and during battle is paramount for the simple reason that, in the words of the Helmuth von Moltke the Elder, chief of staff of the Prussian Army for thirty years, "No plan survives contact with the enemy." Any battlefield is chaotic swarm intelligence in action. For a battle plan to adapt, each agent has to know the purpose of the mission, understand their part in it, have the ability to communicate on the battlefield to update their tactical knowledge of the enemy, coordinate with other agents and adjacent units about their positions and disposition, and make decisions quickly as information deteriorates and change accelerates.

The first element of communication and decision making on the battlefield starts prior to engagement, and it's called the commander's intent:

> The commander's intent describes the desired endstate. It is a concise statement of the purpose of the operation and must be understood two levels below the level of the issuing commander. It must clearly state the purpose of the mission. It is the single unifying focus for all subordinate elements. It is not a summary of the concept of the oper-

ation. Its purpose is to focus subordinates on what has to be accomplished in order to achieve success, even when the plan and concept no longer apply, and to discipline their efforts toward that end.[43]

In the best case the commander's intent is known and understood by all sailors or soldiers so that as the plan deteriorates in battle, individuals can use adaptive behavior to advance the mission. "The commander's intent," suggests Chuck, "should be embodied, embedded, in the warriors." That means that design of the military Evolvabots has to involve Command and Control during design because intelligence and intent, as we've seen throughout this book, are part of not just the programmable nervous system but also the type, arrangement, and quality of sensors, motors, and chassis.

After talking to Chuck I was wondering if the commander's intent (CI) itself could serve as the fitness function for military Evolvabots. Whatever the CI—cause maximum damage to target X or guard squadron Y or rescue fleet Z—the ongoing performance of individual Evolvabots can be judged relative to it. Because the performance of each individual is compared to that of others in its population, the feedback about what works is relevant automatically. It turns out that engineers working for the US Army have already tried, in digital simulation, the idea of using the CI as the fitness function: "Evolution continues until the system produces a final population of high-performance plans which achieve the commander's intent for the mission."[44] Did you get that? Tactical plans, which are extremely complicated themselves, can be evolved using genetic algorithms that use the CI as the fitness function.

The fundamental communication issue on the battlefield is that small groups and isolated individuals have to make decisions without checking with their commanders. As Lieutenant Colonel Lawrence G. Shattuck, professor, Behavior and Sciences and Leadership at the US Military Academy, West Point, has written, the pace of events on the battlefield often precludes direct contact with superiors even if communication channels are open.[45] Once communication ceases, for whatever reason, soldiers need to know the CI to help frame their decisions, getting inside the commander's head to know how she would be making the decision, according to Shattuck.

EVOLVABOTS GET A CONSCIENCE

The central technical challenge is this: get autonomous robots working, communicating, self-repairing, reproducing, and evolving in the wild, without help from humans. Proximal challenges, in addition to the ones already discussed in this chapter, include the following:

1. How do we embed the fitness function in a population of freely roaming Evolvabots?
2. How do we have that fitness function, which is imposed by humans, be an automatic part of the world in which the Evolvabots are working?
3. Or do we let the fitness function be unspecified but emerge from the survival of the robots in the world?
4. In any of these scenarios, how do we monitor and control Evolvabots in the wild?

If these issues can be solved, then everything in this chapter is feasible.

But wait. Just because we can do something, does that mean that we want to—or should? The central ethical challenge, framed by Ronald Arkin in his research on the topic for the US Army Research Office, is this: get autonomous robots to behave "within the bounds prescribed by the Laws of War and Rules of Engagement."[46] "The advent of autonomous robotics in the battlefield," writes Arkin, "as with any new technology, is primarily concerned with *Jus in Bello* [acceptable limits to conduct in war], that is, defining what constitutes the ethical use of these systems during conflict, given military necessity."[47] Arkin's goal—which I support wholeheartedly—is to have our military robots outperform our human soldiers in terms of ethical conduct. Evolving robots need a conscience.

FIGURE 8.4. *So many possible paths, so little retro-futuristic time.* The author points in the conceptual direction that he predicts the field of evolutionary biorobotics will take. He is correct 100 percent of the time.

CLOSING AND OPENING REMARKS—RETROFUTURISM

Predicting the future is even easier than understanding the past. That is the fundamental tenet (tenet number 1), as I see it, of the art movement created by Lloyd Dunn, known as Retrofuturism (Figure 8.4). As you've seen in this chapter, I've been able to predict, with virtually no data to support my arguments, a scenario in which we are at the beginning of a new kind of military arms race. Evolving robots, I claim, will alter the way we fight wars and defend ourselves. For retrofuturistic completeness I should also predict the exact opposite (tenet number 2), namely that evolving robots are a trivial sideshow in the growing field of robotics and have nothing to tell us about the future of warfare.

I don't really think that the second prediction is true. Too bad. In all seriousness, this chapter has been a bummer, right? Who wants to talk about war and autonomous killing machines when we can talk about studying the evolution of the first vertebrates instead? I don't. But the reality is that evolving robots are and will be created for academic, industrial, and military purposes. This means that we should all become students of robots of any kind, whether they be evolving robots, nonevolving autonomous robots, or semiautonomous and remotely controlled military robots. We need to understand robots so we can proceed with due caution and deliberation. No secrets. No surprises.

Now for an apology. In this book I've covered just a tiny sliver of the world of robotics: evolving robots. And I haven't even done that little bit justice. I'm sorry. For example, I've talked mostly about the work done by myself and my collaborators, referring just occasionally and superficially to the great researchers who have inspired us: Ronald Arkin, who is creating the field of robot ethics after unifying behavior-based robotics together; Barbara Webb, who created biorobotics; Stefano Nolfi and Dario Floreano, who created evolutionary robotics; and Valentino Braitenberg and Rodney Brooks, who cocreated the field of behavior-based robotics that jump-started all of the above. To help overcome my guilt for giving all these masters short shrift, I'll tell you that they all have written great books on their subjects, and you should read them.

I should make a parallel apology for the world of evolutionary biology. With a head start of over a hundred years on robotics, evolutionary biology and my omissions of name are more difficult to characterize and recognize. I can tell you that I've largely ignored the fascinating world of EvoDevo, the evolution of ontogenies and the constraints and possibilities that developmental systems give to the species they construct. Sean Carroll is the place to start reading there. Vertebrate paleontologists like David Raup, Steven Stanley, Robert Carroll, and Michael Benton ought to feel slighted because they have carried the torch and blazed the trail with their excellent textbooks. The great biomechanicists, McNeil Alexander, Tom Daniel, John Gosline, Mark Denny, Paul Webb, Andy Biewener, get nary a mention. You have to leave out a lot, I've learned, when you write a book.

And finally, here's a parting shot, a reminder of one of the concepts that I consider most important and most often misunderstood:

evolution. I've been using the word "evolution" throughout this book *not* in its causal, on-the-street sense of directed, progressive, optimizing design but rather in its scientific sense: a change in a population of agents over generational time, in a given environment, caused by random genetic processes coupled with selection, where selection results from the interaction of individual agents with and within their physical environments.

We've talked about, designed, experimented with, and analyzed the behavior of two types of robots: (1) Evolvabots, which change in ways that we can't predict with certainty when they constitute a population with genetic inheritance and selection pressure; (2) Evolutionary Trekkers, which we create to be of a certain form in order to help us understand how extinct or never-existing forms may have behaved in a given environment. Both Evolvabots and ETs are built to test ideas about evolution as a process (how things can evolve), specific evolutionary events (selection pressures), or specific evolutionary situations (big, four-flippered animals).

Although I've focused on fish and aquatic vertebrates in this book, just keep in mind that you can use Evolvabots and ETs to model any kind of critter.

So long, and thanks for all the robotic fish.

ACKNOWLEDGMENTS

This book exists because of the creativity, hard work, and acumen of three terrific professionals. Jeff Kosmacher, director of media relations and public affairs at Vassar College; my agent, Laura Wood, of FinePrint Literary Management; and my editor, T. J. Kelleher of Basic Books. Jeff got Mike Hill, of the Associated Press, to report on our evolving robots. Laura read Mike's story, conceived of a book, convinced me to write it, and then found it a wonderful home at Basic Books. T. J. gave me the freedom to write what I wished and then skillfully reworked the manuscript into a book. My respect and appreciation for this trio are infinite.

My thanks to the team at Basic Books who provided invaluable support: Sandra Beris, project editor; Josephine Mariea, copyeditor; Tisse Takagi, associate editor; Michele Jacob, VP and director of publicity; and Andrea Bussell, publicity manager.

Two environments have allowed me to thrive: home and work. Meg, Isabel, Adele, Madeleine, Tamasin, Capercaillie, and Kookaburra have made family life bustling, bright, and restorative. Karabadangbaraka! In the Biology Department at Vassar I've been lucky to have generous and inspiring colleagues: Jerry Calvin, Lynn Christenson, Tebbie Collins, Pauline Contelmo, Erica Crespi, Jeremy Davis, Kelli Duncan, David Esteban, Dick Hemmes, Barbara Holloway, David Jemiolo, Jason Jones, Jenni Kennell, Betsy Ketcham, Sue Lerner, Ann Mehaffey, Leathem Mehaffey, Bonnie Milne, Sue

Painter, Marshall Pregnall, Mark Schlessman, Bill Straus, Kate Susman, Bob Suter, Nancy Pokrywka, Jodi Schwarz, Lina Spallone, Julie Williams, and Keri VanCamp. In the Cognitive Science Program Jan Andrews, Gwen Broude, Carol Christensen, Kathleen Hart, Ken Livingston, and Carolyn Palmer have all been adventurous enough to team-teach with me and, in so doing, serve as my tutors in the vast multidisciplinary field that is cognitive science.

The research in this book is the joyous consequence of working with brilliant, happy collaborators, folks whom Laura would call "fun-loving nerds who want to learn something cool." Six have been fool enough to work with me continually for the past ten to twenty years: Tom Koob (MiMedx Group, Inc.), Chun Wai Liew (Lafayette College), Ken Livingston (Vassar), Matt McHenry (University of California, Irvine), Charles Pell (Physcient, Inc.), and Robert Root (Lafayette College). Others showing questionable judgment in their professional affiliations include, at Vassar College, Carl Bertsche, Peter Czuwala, Larry Doe, Tom Ellman, Luke Hunsberger, Barbara Holloway, Betsy Ketcham, Jason Jones, Josh de Leeuw, Nick Livingston, Marianne Porter, Bradley Richards, Bob Suter, and John Vanderlee; Melina Hale and Mark Westneat at the University of Chicago and the Field Museum; Ann Pabst and Bill McLellan at the University of North Carolina, Wilmington; Matthieu Kemp and Brett Hobson at Nekton Research, Inc.; Farshad Khorrami and Prashanth Krishnamurthy at Farco Technologies, Inc. and New York University; Miriam Ashley-Ross (Wake Forest University), Barbara Block (Stanford University), Hugh Crenshaw (Physcient, Inc.), Shelley Etnier (Butler University), Randy Ewoldt (University of Illinois at Urbana-Champaign), Frank Fish (West Chester University), Alice Gibb (Northern Arizona University), Sindre Grotmol (University of Bergen), Mary and Jack Hebrank (Duke University and North Carolina State University), Lena Koob-Emunds (Mount Desert Island Biological Laboratory), Doug Pringle (MiMedx Group, Inc.), Fred Schachet (Duke University), Justin Schaefer (University of California, Los Angeles), William "Bart" Shepherd (Steinhart Aquarium), Jim

Strother (University of California, Irvine), Adam Summers (University of Washington), and Phil Watts (Applied Fluids Engineering, Inc).

Students have suffered as my minions and coauthors in the BARK (Biomechanics Advanced Research Kitchen), the IRRL (Interdisciplinary Robotics Research Laboratory), the Abyss (Biorobotics and Biomechanics Laboratory), and the Pirate Republic of Pelagia: Kurt Bantilan, Nick Boetticher, Craig Blanchette, Sasha Cavanagh, Annabeth Carroll, Keon Combie, Hayden-William Courtland, Pam Cuce, Megan Cummins, Candido Diaz, Nicole Doorly, Gigi Engel, Carina Frias, Andres Gutierrez, Jonathan Hirokawa, Kira Irving, McKenzie Johnson, Wyatt Korff, Nicole Krenitsky, Nicole Lamb, Adam Lammert, Gianna McArthur, Karen Nipper, David Paul, Sonia Roberts, Greg Rodebaugh, Oren Rosenberg, Hannah Rosenblum, Hassan Sakhtah, Jonathan Seclow, Joe Schumacher, Avery Siciliano, Ben Sinwell, Elise Stickles, Adina Suss, Joshua Sturm, Eamon Twohig, and Stephanie Varga.

My heroes are my teachers. In the department of zoology at Duke University, Professors Steve Wainwright and Steve Vogel, the fathers of comparative biomechanics, taught me the ropes. Professors Fred Nijhout, John Lundberg, Steve Nowicki, Louise Roth, Kathleen Smith, and Vance Tucker deserve part of the blame, as do senior graduate students in the BLIMP and physiology groups: Barbara Block, Hugh Crenshaw, Olaf Ellers, Matt Healy, Anne Moore, Lisa Orton, and Ann Pabst. I wouldn't have gotten to Duke without Sentiel "Butch" Rommel (College of the Atlantic) and Ted Grand (Smithsonian National Zoological Park), both of whom taught me that deep, genuine enthusiasm for any subject was the spark that lit the mind's fire.

Friends and family offered encouragement and editorial advice: Paul Callagy and Carolyn Palmer, Mary Ann Cunningham and Tom Finkle, Doug Eaton, Keith and Lisa Fadelici, Kate and Byron Jordan, John Keller, Daniel and Ken Lockhart, Dale Long, Jeb Long, Kate and Joel Nimety, Marty Ronsheim, Jody and Paul Ronsheim, Amy and Lloyd Spencer, Ann and Jeff Staten, Sharon Swartz, Tracy and Joe Troy. Jeff Staten offered trenchant insights in Chapter 1, Adam Lammert was of great help framing Chapter 5, and Charles Pell and Ted Grand made important contributions to Chapter 8.

Some material in this book is based on work supported by the National Science Foundation (grant numbers IOS-0922605, DBI-0442269, BCS-0320764, IOS-9817134). Any opinions, findings, and conclusions or recommendations expressed in this book are those of the author and do not necessarily reflect the views of the National Science Foundation.

Some material in this book is based on work supported by the Department of the Navy, Office of Naval Research (grant numbers N00014097-1-0292, N00014-93-1-0594). Any opinions, findings, and conclusions or recommendations expressed in this book are those of the author and do not necessarily reflect the views of the Office of Naval Research.

NOTES

CHAPTER 2

1. A cautionary note: I'm being flippant here in several ways. First, the list of behaviors in this paragraph is vertebrate-centric. Second, it's unlikely that any individual vertebrate alters behavior very much in its own lifetime; big changes in behavioral strategy occur over generational time.

2. Scoring points by helping your relatives raise kin is formally recognized as "inclusive fitness."

3. Although it probably seems crazy at first, I'm not sure that evolving robots aren't alive. The quality that we recognize scientifically as "life" is usually a suite of characteristics that include the ability of a self-contained entity to (1) make additional and similar versions of itself, (2) gather energy, (3) convert and use the gathered energy to perform chemical and mechanical work (e.g., build or repair itself, gather more energy, make copies of itself), and (4) decrease entropy (disorder) locally and temporarily. These ideas are influenced, in part, by the physicist Erwin Schrödinger in his 1944 book, *What Is Life?* (based on lectures delivered under the auspices of the Dublin Institute for Advanced Studies, Dublin, February 1943). From cognitive science, I would include, in our life list, the ability of a self-contained entity to (5) react to changes in the patterns of the global energy array with the goal of reproducing and gathering energy. What do you think?

4. This definition of natural selection is a bit different from that found in textbooks. In Mark Ridley's excellent textbook, *Evolution* (3rd ed.)(Malden, MA: Blackwell Science, 2004), he enumerates Darwin's four necessary and sufficient conditions for natural selection: (1) reproduction, (2) heredity, specifically, offspring resembling parents, (3) variation in individual characters among members of the population, and (4) that variation in reproductive output for individuals is tied to the variation in characters. Note here that the concept of a population is secondary; I make it primary in order to emphasize the concept of interaction of individuals and their world being defined by who they are with and where they are.

5. You can take this journey by reading Jonathan Weiner's book *The Beak of the Finch: A Story of Evolution in Our Time* (New York: Alfred A. Knopf, 1994).

6. This insight—that slow, small changes happening right in front of us are sufficient to drive large-scale and dramatic changes over long periods of time—is the basis for explaining any kind of evolutionary change. When we've published evolutionary simulations with robots that measure change over generational time, we claim that we are learning something about the evolutionary processes that have occurred over millions of years to help create new traits and new species.

7. John Stuart Mill created methods for inferring causal relations that includes *ceteris paribus*. Mill's methods and other reasoning techniques are explained in an accessible manner in David Kelley's *The Art of Reasoning, 3rd ed.* (New York: W. W. Norton & Company, 1998).

8. Robert Brandon, *Adaptation and Environment* (Princeton, NJ: Princeton University Press, 1990).

9. Barbara Webb, "Can Robots Make Good Models of Biological Behaviour?" *Behavioral and Brain Sciences* 24, no. 6 (2001): 1033–1055.

CHAPTER 3

1. Richard Feynman, the Nobel Laureate in Physics, said something similar: "What I cannot create, I do not understand."

2. What's really useful about the secret code is that it implies that if you can't build it (whatever "it" is), then you don't really understand it. This is another way of thinking about what people call an "existence proof," which is the ultimate in physical evidence: if it exists, then it can exist.

3. For an in-depth examination of representation, see Tim Crane's *Mechanical Mind: A Philosophical Introduction to Minds, Machines, and Mental Representation, 2nd ed.* (London, New York: Routledge, 2003).

4. In other words, make a plan. Once you have an explicit plan researched and written down—the answers to the six design questions are a good start—then keep in mind what General Dwight Eisenhower said: "In preparing for battle I've always found that plans are useless, but planning is indispensible."

5. If you are interested, these issues are addressed in the fast-paced fields of phylogenetics, phylogenomics, and evolutionary developmental biology.

6. In the mid-naughties—2000s—the flux in our understanding of evolutionary relationships was widely recognized in college-level textbooks. For an excellent example, see Chapter 1 in Michael J. Benton, *Vertebrate Paleontology, 3rd ed.* (Malden, MA: Blackwell Science, 2005). Also, Frédéric Delsuc, Henner Brinkmann, Daniel Chourrout, and Hervé Philippe, "Tunicates and Not Cephalochordates Are the Closest Living Relatives of Vertebrates," *Nature* 439, no. 7079 (2006): 965–968. This then-radical view has been tentatively accepted (but still subject to revision as new data are generated), particularly in light of new information on the genome of cephalochordates. See Peter W. H. Holland, "From Genomes to Morphology: A View from Amphioxus," *Acta Zoologica* 91, no. 1 (2010): 81–86.

7. Tunicates (also called ascidians and urochordates) are members of the Phylum Chordata, a group of related species that also contains vertebrates and lancelets. Lancelets, also called amphioxus (species name *Branchiostoma*), are members of the Cephalochordata. For an excellent comparison of these two groups in the context of vertebrate origins, see M. Schubert, H. Escriva, J. Xavier-Neto, and V. Laudet, "Amphioxus and Tunicates as Evolutionary Model Systems," *Trends in Ecology and Evolution* 21, no. 5 (2006): 269–277.

8. The phrase "ugly bags of mostly water" was a first-contact description of humanoids uttered by a crystalline life-form in the *Star Trek: The Next Generation* episode "Home Soil" (season 1). If these crystalline life-forms had seen adult tunicates and humanoids side by side, I'm guessing that they would've called the former bags and the latter, more accurately, tubes or branched cylinders.

9. Walter Garstang, "The Morphology of the Tunicata, and Its Bearing on the Phylogeny of the Chordata," *Quarterly Journal of Microscopical Science* 62 (1928): 51–187. In one of those beautiful, accidental collusions between science and art, his scientific ideas are immortalized in his poetry, published posthumously in 1951 as *Larval Forms and Other Zoological Verses* (Oxford: Blackwell).

10. You can read about Tadro1 in J. H. Long Jr., C. Lammert, C. A. Pell, M. Kemp, J. Strother, H. C. Crenshaw, and M. J. McHenry, "A Navigational Primitive: Biorobotic Implementation of Cycloptic Helical Klinotaxis in Planar Motion," *IEEE Journal of Oceanic Engineering* 29, no. 3 (2004): 795–806.

11. J. H. Long Jr., T. J. Koob, K. Irving, K. Combie, V. Engel, N. Livingston, A. Lammert, and J. Schumacher, "Biomimetic Evolutionary Analysis: Testing the Adaptive Value of Vertebrate Tail Stiffness in Autonomous Swimming Robots," *Journal of Experimental Biology* 209, no. 23 (2006): 4732–4746.

12. You can see the original pictures of the fossils in this paper: D.-G. Shu, S. Conway Morris, J. Han, Z.-F. Zhang, K. Yasui, P. Janvier, L. Chen, X.-L. Zhang, J.-N. Liu, Y. Li, and H.-Q. Lui, "Head and Backbone of the Early Cambrian Vertebrate *Haikouichthys*," *Nature* 421, no. 6922 (January 2003): 526–529.

13. Sindre Grotmol, Harald Hryvi, Roger Keynes, Christel Krossøy, Kari Nordvik, and Geir K. Totland, "Stepwise Enforcement of the Notochord and Its Intersection with the Myoseptum: An Evolutionary Path Leading to Development of the Vertebra?" *Journal of Anatomy* 209, no. 3 (2006): 339–357.

14. "Stiffness" by itself is an ambiguous term. Many different kinds of stiffness exist. What they have in common is that they are all a proportionality constant between an applied force, stress, or torque and the resulting change in length, strain, or curvature, respectively. Flexural stiffness is defined as the proportionality constant between a torque and the resulting curvature. Flexural stiffness assumes that this relationship is the same all along the length of the structure that you are bending. If you want to describe the stiffness of the whole structure, a physicist would talk about "spring constant" or a "spring stiffness." I prefer the term "structural stiffness." In the case of a cantilevered beam with a weight hung off the end, the structural stiffness is the ratio of the flexural stiffness to the cube of the beam's length.

This means that you can have beams of constant flexural stiffness but different structural stiffness if you vary their lengths.

15. For reconstructions of the skeletons and their partial vertebrae, see J. A. Long and M. S. Gordon, "The Greatest Step in Vertebrate History: A Paleobiological Review of the Fish-Tetrapod Transition," *Physiological and Biochemical Zoology* 77, no. 5 (2004): 700–719.

16. The use of the term "biomimetic" varies across engineering, bioengineering, and biomedical engineering. Here I use "biomimetic" to mean a system built to resemble, as much as possible, the biological target.

17. Rolf Pfeifer, and Christian Scheier, *Understanding Intelligence* (Cambridge, MA: MIT Press, 1999).

18. Hou Xian-Guang, Richard J. Aldridge, Jan Bergstron, David J. Siveter, Derek J. Siveter, and Feng Xiang-Hong, *The Cambrian Fossils of Chengjiang, China: The Flowering of Early Animal Life* (Malden, MA: Wiley-Blackwell, 2007).

19. Yes, I'm referencing the television show *Survivor* on CBS. Their logo reads, "Outwit, outplay, outlast." I in no way mean to imply that reproduction is or should be part of this show.

20. Cameron K. Ghalambor, Jeffrey A. Walker, and David N. Reznick, "Multitrait Selection, Adaptation, and Constraints on the Evolution of Burst Swimming Performance," *Integrative and Comparative Biology* 43, no. 3 (2003): 431–438. Also see R. B. Langerhans, "Predicting Evolution with Generalized Models of Divergent Selection: A Case Study with Poeciliid Fish," *Integrative and Comparative Biology* 50, no. 6 (2010): 1167–1184.

21. Barbara Webb, "Can Robots Make Good Models of Biological Behaviour?" *Behavioural and Brain Sciences* 24, no. 6 (2001): 1033–1050. Also see Webb, "Validating Biorobotic Models," *Journal of Neural Engineering* 3, no. 3 (September 2006): R25–R35. I use my rephrased terms of Webb's dimensions in my paper "Biomimetic Robotics: Self-Propelled Physical Models Test Hypotheses about the Mechanics and Evolution of Swimming Vertebrates," *Proceedings of the Institution of Mechanical Engineers, Part C, Journal of Mechanical Engineering Science* 221, no. 10 (2007): 1193–1200.

CHAPTER 4

1. Plagiarism detector alert: a thank you to Lewis Carroll.

2. This single number that measures feeding behavior is a composite from the fitness function that we developed in Chapter 3. We defined better "relative fitness" for an individual in a given generation, relative to other individuals in that generation, as the sum of their scaled values for increased swimming speed, decreased time to the light target, reduced distance from the light target over the course of the whole experiment, and reduced wobble as they moved. These relative fitness values only make sense within a generation relative to other competing individuals at that time and place: they can't be compared across generations. To make those cross-

generation comparisons for Figure 4.1, we compared any individual's performance to the average of all individuals over all ten generations scaled by the standard deviation of the particular sub-behaviors, speed, time, distance, and wobble. In statistical terms, we summed up the z-scores of each sub-behavior for each individual.

3. In statistics one standard deviation, which changes in value depending on the situation, is a measure of how far away from the average most numbers in a group of numbers fall. A small standard deviation means that most numbers in the group are close to the average of the group.

4. If you are interested in the mathematics of the mating that we used, you can find the details in our paper on the evolution of Tadro3: J. H. Long Jr., T. J. Koob, K. Irving, K. Combie, V. Engel, N. Livingston, A. Lammert, and J. Schumacher, "Biomimetic Evolutionary Analysis: Testing the Adaptive Value of Vertebrate Tail Stiffness in Autonomous Swimming Robots," *Journal of Experimental Biology* 209, no. 23 (December 2006): 4732–4746.

5. In John Gillespie's *Population Genetics: A Concise Guide, 2nd ed.* (Baltimore: Johns Hopkins University Press, 2004), he speaks of "demographic stochasticity" as this source of small-number randomness. He also points out a second such source, the segregation of the different parental alleles into separate gametes. Both sources together he calls genetic drift. In our robotic simulation segregation is not a factor because our quantitative characters are, by design, split evenly between chromosomes.

6. Students and scholars of evolutionary theory will be quick to interject, what about sexual selection, gene flow, genetic drift, epistasis, mating, and developmental processes as evolutionary mechanisms? True. Those are other identifiable mechanisms of evolutionary change. Lenski's point, which I follow here, is that any mechanism fits into a category of either being deterministic or random. Natural selection is deterministic in that once you identify all of Brandon's information (see Chapter 2), you can predict evolutionary outcome. Random factors like mutation or assortative mating have outcomes that are not predictable. I continue to be influenced by this fascinating and illuminating paper: M. Travisano, J. A. Mongold, F. Bennett, and R. E. Lenski, "Experimental Tests of the Roles of Adaptation, Chance, and History in Evolution," *Science* 267, no. 5194 (1995): 87–90. Also, you may be interested in the Neutral Theory of molecular evolution, which is based on the idea that most random genetic changes have no effect on selection. In the face of genomic data, this idea is rapidly changing: Matthew W. Hahn, "Toward a Selection Theory of Molecular Evolution," *Evolution* 62, no. 2 (2007): 255–265.

7. For completeness, I should tell you that the variable I (in units of meters to the fourth power) is called the "second moment of area." The second moment of area is a geometric property of how the structure's material is arranged and clustered in cross-section, the plane perpendicular, to, in this case, the long axis of our beam that we measure with the variable L.

8. You can find a great introduction to the evidence for our mental modeling in the following book: Read Montague, *Your Brain Is (Almost) Perfect: How We Make Decisions* (New York: Plume, 2006).

9. The philosopher of science, Karl Popper, has formalized the "hypothetical-deductive" methodology in order to avoid what other philosophers have called the "problem of induction," or generalizing from a few observations to the world in general. Most of our statistical hypothesis testing in science is structured around the idea of falsification or rejection of the "null" hypothesis. The danger with this approach is that if you reject the null hypothesis, you are tempted to treat the alternative as "true," when in fact it becomes the new null to be tested. An excellent place to start with this kind of careful inference is with Popper himself: Karl R. Popper, *The Logic of Scientific Discovery* (New York: Basic Books, 1959).

10. For that matter (ahem . . .), no one has seen energy. In fact, physicists don't even know what energy is. Richard Feynman, the Nobel Laureate in Physics, and his coauthors Robert Leighton and Matthew Sands point this out eloquently in *The Feynman Lectures on Physics, vol. 1* (Reading, MA: Addison-Wesley, 1964), 4-2.

11. I've given short shrift here to an interesting philosophical debate: logical positivism versus Popper's hypothetico-deductivism. One distillation of the difference is *modus ponens* versus *modus tollens* logic, respectively.

12. In case you are interested in what we did to try to find our flaws, here's an example. We were very concerned that our initial measurements of structural stiffness were somehow flawed. We had created a standard curve that gave us a value of material stiffness, E, for a given amount of gelatin and cross-linking time. We retested that formula to make sure it was accurate. Moreover, if our method for making and measuring the biomimetic notochords was highly variable, that would be an additional source of random variation and noise. To test this we split up into three different groups and completely remade all of our tails, and then we tested them for structural stiffness using a materials testing device. We compared the three groups for inter-rater reliability, the level of agreement between us. In the worst case the correlation of our stiffness measurements between groups was 0.91 out of 1.0.

13. In statistics-ese these least-squares linear regressions are all "highly significant," with $p < 0.01$ in each case. Prior to testing, all data were transformed so that they fit a normal distribution. The 20 percent refers to the "coefficient of determination," also called the "r-squared value," a number from 0 to 1 that indicates how well the best-fit line represents the relationship between the dependent and independent variables.

14. Warning: "epistasis" has different meanings. For example, Gillespie (*Population Genetics*) defines three different kinds of genetic epistasis. Here I am taking the broader view of interactions among genes impacting fitness, similar to "functional epistasis." I'm interested in the gene-to-fitness mapping mediated by phenotype. In Tadro3 wobble and speed interact. Because both are correlated with stiffness, and stiffness is genetically coded, selection on wobble and speed alter the genetics of the population. Simulations show that epistatic networks can adapt: Roman Yukelevich, Joseph Lachance, Fumio Aoki, and John R. True, "Long-Term Adaptation of Epistatic Genetic Networks," *Evolution* 62, no. 9 (2008): 2215–2235.

15. In case your bad-grammar detector has signaled, I should explain that I'm trying to make a self-referential joke.

16. Cameron K. Ghalambor, Jeffrey A. Walker, and David N. Reznick, "Selection, Adaptation and Constraints on the Evolution of Burst Swimming Performance," *Integrative and Comparative Biology* 43, no. 3 (2003): 431–438.

17. Rowan D. H. Barrett, Sean M. Rogers, and Dolph Schluter, "Natural Selection on a Major Armor Gene in Threespine Stickleback," *Science* 322, no. 5899 (2008): 255–257.

18. Richard W. Blob, Sandy M. Kawino, Kristine N. Moody, William C. Bridges, Takashi Maie, Margaret B. Ptacek, Matthew L. Julius, and Heiko L. Schoenfuss, "Morphological Selection and the Evaluation of Potential Tradeoffs Between Escape from Predators and the Climbing of Waterfalls in the Hawaiian Stream Goby *Sicyopterus Stimpsoni*," *Integrative and Comparative Biology* 50, no. 6 (2010): 1185–1199, doi:10.1093/icb/icq070.

CHAPTER 5

1. This ability to "know" or infer if any other agent, organic or artificial, possesses a mind is an exciting area of philosophical and scientific work that's usually called, "The Problem of Other Minds." I realize that I'm conflating "mind" and "intelligence" here. The two are often treated as interchangeable: a human mind is considered, by definition, intelligent; thus, so it goes for some, intelligence is found only in human minds.

2. Interaction of a human and a potential artificial intelligence is the basis of what Turing called the "imitation game." We now call this the "Turing Test." Read all about it in his wonderfully accessible paper: Alan Turing, "Computing Machinery and Intelligence," *Mind* 59, no. 1 (1950): 433–460.

3. Here's the official site of the Loebner Prize: www.loebner.net/Prizef/loebner-prize.html.

4. Stevan Harnad, "The Turing Test Is Not a Trick: Turing Indistinguishability Is a Scientific Criterion," *SIGART Bulletin* 3, no. 4 (October 1992): 9–10.

5. Impossible as the T^3 may seem to us as we contemplate human-level performance, I argue that the T^3 has been passed, perhaps even at the level of the Loebner Prize gold medal, for a different species: cockroaches. Autonomous cockroach robots fooled real cockroaches so well that they could cause the real cockroaches to do things they didn't normally do, like form groups in the light (they prefer the dark). Here's the brilliant paper: J. Halloy et al., "Social Integration of Robots into Groups of Cockroaches to Control Self-Organized Choices," *Science* 318, no. 5853 (November 2007): 1155–1158.

6. This paper contains an excellent description of Searle's classic "Chinese room" thought experiment: John Searle "Is the Brain's Mind a Computer Program?" *Scientific American* 202, no. 1 (1990): 26–31.

7. Experimental evidence for self-recognition in dolphins can be found in this paper: D. Reise and L. Marino, "Mirror Self-Recognition in the Bottle-Nose Dolphin: A Case of Cognitive Convergence," *Proceedings of the National Academy of Sciences* 98, no. 10 (2001): 5937–5942.

8. H. M. Gray, K. Gray, and D. M. Wegner, "Dimensions of Mind Perception," *Science* 315 (2007): 619.

9. In 1982 Vassar College became the first institution in the world to offer an undergraduate major in cognitive science. Hampshire College disputes Vassar's claim of primacy (see the claim on the website of their School of Cognitive Science at www.hampshire.edu/cs/).

10. A terrific exploration of embodied intelligence is Louise Barrett's *Beyond the Brain: How Body and Environment Shape Animal and Human Minds* (Princeton, NJ: Princeton University Press, 2011).

11. The particular microcontroller we used for Tadros 2 to 4 was a HandyBoard, invented by Fred Martin at MIT (see en.wikipedia.org/wiki/Handyboard for a great picture and a useful overview). The original Tadro1 had a completely analog electronic brain.

12. The software that runs on the Handyboard, Interactive C, was originally developed for LEGO Robotics competitions. Two versions of Interactive C are available, one by Newton Labs (www.newtonlabs.com/ic/) and the other by the KISS Institute (www.botball.org/ic).

13. Alva Noë, *Action in Perception* (Cambridge, MA: MIT Press, 2004).

14. For a review of the work of Floreano and his colleagues on this topic, see Mototaka Suzuki and Dario Floreano, "Enactive Robot Vision," *Adaptive Behavior* 16, nos. 2–3 (2008): 122–128. Enactive perception has also been put to good use in the training of AMAR-III, a humanoid robot that categorizes objects based on its enactive experience with them (http://spectrum.ieee.org/robotics/artificial-intelligence/a-robots-body-of-knowledge).

15. George Lakoff and Mark Johnson, *Philosophy in the Flesh: The Embodied Mind and Its Challenge to Western Thought* (New York: Basic Books, 1999).

16. Ecological psychology was created by J. J. Gibson. Here's a great place to start: J. J. Gibson, "Visually Controlled Locomotion and Visual Orientation in Animals," *British Journal of Psychology* 49, no. 3 (1958): 182–194.

17. Lawrence W. Barsalou, "Grounded Cognition," *Annual Review of Psychology* 59 (2008): 617–645.

18. Once again I refer you to Pfeifer and Scheier's excellent book, *Understanding Intelligence*.

19. Even if you have heard about Phineas Gage before, you should read this fascinating paper: H. Damasio, T. Grabowski, R. Frank, A. M. Galaburda, and A. R. Damasio, "The Return of Phineas Gage: Clues about the Brain from the Skull of a Famous Patient," *Science* 264, no. 5162 (1994): 1102–1105.

20. NOVA, "Musical Minds," www.pbs.org/wgbh/nova/musicminds/. A video fMRI of Sack's brain listening to music can be found at www.pbs.org/wgbh/nova/musicminds/extra.html.

21. J. M. Fuster, "Upper Processing Stages of the Perception-Action Cycle," *Trends in Cognitive Science* 8, no. 4 (2004): 143–145.

22. E. Tytell, C-Y. Hsui, T. L. Williams, A. V. Cohen, and L. Fauci, "Interactions Between Internal Forces, Body Stiffness and Fluid Environment in a Neurome-

chanical Model of Lamprey Swimming," *Proceedings of the National Academy of Sciences* 107, no. 46 (2010): 19832–19837.

23. Alan Mathison Turing, "Computing Machinery and Intelligence," *Mind* 59, no. 236 (1950): 433–460.

24. Dedre Gentner and B. Bowdle "Metaphor as Structure-Mapping," in *The Cambridge Handbook of Metaphor and Thought*, edited by Raymond W. Gibbs Jr., 109–128 (New York: Cambridge University Press, 2008).

25. David Kelley, *The Art of Reasoning, 3rd ed.* (New York: W. W. Norton & Company, 1998).

26. A good introduction to functionalism and other issues in the philosophy of mind can be found in K. T. Maslin, *An Introduction to the Philosophy of Mind, 2nd ed.* (Malden, MA: Polity Press, 2007).

27. Many different flavors of functionalism exist. "Artificial Intelligence" functionalism, for example, attends to the creation of the same kind of intelligence in vertebrates, computers, and robots. I think you can also make a case for what I call Biological Functionalism: when independent evolutionary events converge on similar functional designs. For example, the brains of birds and mammals are very different in terms of how the different parts have evolved compared to their hypothetical common ancestor. But some birds, like crows and parrots, manage to have brains that allow them to make and use tools and language. These convergent abilities show that some birds and mammals possess the same kind of intelligence (= similar function) that is created by different structures. For a great review of the functional similarities and structural differences of birds and mammals, I recommend the following paper: Ann B. Butler and Rodney M. J. Cotterill, "Mammalian and Avian Neuroanatomy and the Question of Consciousness in Birds" *Biological Bulletin* 211, no. 2 (2006): 106–127.

28. Robert M. Pirsig, *Zen and the Art of Motorcycle Maintenance: An Inquiry into Values* (New York: William Morrow, 1974).

29. Yes, this is the same conceptual cat that Norbert Weiner used, as we spoke of in Chapter 1, to assert problems with modeling: the best model of a cat is a cat. Interesting, is it not, that Weiner's cat surfaces here as a warning about studying the cat itself? The cat has Weiner's tongue.

30. Robert C. Brusca and Gary J. Brusca, *Invertebrates, 2nd ed.* (Sunderland, MA: Sinauer Associates, 2003).

31. Georg F. Striedter, *Principles of Brain Evolution* (Sunderland, MA: Sinauer Associates, 2005).

32. Howard Gardner, *Intelligence Reframed: Multiple Intelligences for the 21st Century* (New York: Basic Books, 1999).

33. This is why, by the way, we do spend time on neuroscience in *Introduction to Cognitive Science* (Cogs 100)! We also spend time on philosophy because of the important attempts to rationally define and understand what in the world we are trying to talk about when we use words and concepts such as mind, brain, behavior, and intelligence.

34. A great place to start is with this book: Patricia Churchland, *Brain-Wise: Studies in Neurophilosophy* (Cambridge, MA: MIT Press, 2002).

35. If you care about building intelligent machines, you must read this book: Jeff Hawkins and Sandra Blakeslee, *On Intelligence: How a New Understanding of the Brain Will Lead to the Creation of Truly Intelligent Machines* (New York: Times Books, 2004).

36. Steven Vogel and Stephen A. Wainwright, *A Functional Bestiary: Laboratory Studies about Living Systems.* (Reading, MA: Addison-Wesley, 1969), 93.

37. Descartes is often considered to be the father of cognitive science because he approached the mind-body problem rationally and scientifically. Even though substance dualism was quickly, even in his day, refuted as a scientific theory, we talk about it in cognitive science because it underwrites so much of our intuition about minds, souls, ghosts, and heaven. For an introduction to dualism, visit the Stanford Encyclopedia of Philosophy: plato.stanford.edu/entries/dualism/#SubDua, or read chapters 1 and 2 in Maslin, *An Introduction to the Philosophy of Mind.*

38. This paper does a great job explaining neural circuits and their functions: D. W. Tank, "What Details of Neural Circuits Matter?" *Seminars in the Neurosciences* 1 (1989): 67–79.

39. Talk here of circuits and what's necessary and sufficient to show causal relations of neural circuits to behavior are largely drawn from his book, which I strongly recommend: Thomas J. Carew, *Behavioral Neurobiology: The Cellular Organization of Natural Behavior* (Sunderland, MA: Sinaur and Associates, 2000).

40. Carew, again. Ibid.

41. I really hate the use of the word "simple" in a context like this because it comes packed with so much anthropocentric baggage. One item in our luggage is that we humans assume that we are the most "complex" organisms by any measure. But consider a single-celled organism: in one cell it packs all the basic functions—like eating, moving, and reproducing—that we humans need a multicellular body to perform. As we go on, you'll see that I call Tadro3 "simple" with specific reference to its sensory-motor system. That's okay, I'd argue, because I'm being explicit about the system of comparison. Implicit "simplicity" means a thousand different, unspoken things.

42. Lakoff and Johnson, *Philosophy in the Flesh.*

43. George Lakoff, "The Neural Theory of Metaphor," in Gibbs, *The Cambridge Handbook of Metaphor and Thought,* 17–38.

44. Louise Barrett, *Beyond the Brain: How Body and Environment Shape Human Minds* (Princeton, NJ: Princeton University Press, 2011).

45. Neurocomputational modeling of swimming vertebrates by Örjan Ekeberg and then Auke Ijspeert have shown that many, many possible circuit structures will produce the same function (functionalism rules!). Thus, we shouldn't take the two T3 circuits here as the only ones that are possible. Örjan Ekeberg, "A Combined Neuronal and Mechanical Model of Fish Swimming," *Biological Cybernetics* 69, nos. 5–6 (1993): 363–374. Auke Jan Ijspeert, John Hallam, and David Willshaw, "Evolving Swimming Controllers for a Simulated Lamprey with Inspiration from Neurobiology," *Adaptive Behavior* 7, pt. 2 (1999): 151–172.

46. Valentino Braitenberg, *Vehicles: Experiments in Synthetic Psychology* (Cambridge, MA: MIT Press, 1984), 20.

47. Ibid.

48. Brooks chronicles this revolution: Rodney A. Brooks, *Flesh and Machines: How Robots Will Change Us* (New York: Pantheon Books, 2002).

49. For more on Brooks's Ghenghis, including its ancestors and descendants, dig around at the website of the MIT Computer Science and Artificial Intelligence Laboratory: www.csail.mit.edu/.

50. Behavior-based robotics, as a field, was codified by Professor Ronald Arkin in his seminal textbook, *Behavior-Based Robotics* (Cambridge, MA: MIT Press, 1998). Behavior-based robotics is now recognized as one of the first successful forays into the general field of biologically inspired artificial intelligence. For more, see Dario Floreano and Claudio Mattiussi, *Bio-Inspired Artificial Intelligence: Theories, Methods, and Technologies.* (Cambridge, MA: MIT Press, 2008).

51. Rodney Brooks, "A Robust Layered Control System for a Mobile Robot," A.I. Memo 864, MIT Artificial Intelligence Laboratory, 1985. Published as R. Brooks, "A Robust Layered Control System for a Mobile Robot," *IEEE Journal of Robotics and Automation* 2, no. 1 (1986): 14–23.

52. Matt McHenry, whom you may remember as one of the inventors of Tadro1, and his PhD student, William Stewart, combined experiments on and models of zebrafish to look at the possible influence of a predator on flow around a prey: W. J. Stewart, and M. J. McHenry, "Sensing the Strike of a Predatory Fish Depends on the Specific Gravity of a Prey Fish," *The Journal of Experimental Biology* 213, pt. 22 (November 2010): 3769–3777.

53. For a review of fast starts in the context of predatory-prey situations, read: P. Domenici, "Scaling of Locomotor Performance in Predatory-Prey Encounters: From Fish to Killer Whales," *Comparative Biochemistry and Physiology Part A* 131 (2001): 169–182.

54. This figure of 3 G during takeoff is from mission specialist Koichi Wakata and can be found at the NASA website: spaceflight.nasa.gov/feedback/expert/answer/crew/sts-92/index.html.

55. For a lovely review of this fast-start escape circuit, see the following papers: S. J. Zottoli and D. S. Faber, "The Mauthner Cell: What Has It Taught Us?" *The Neuroscientist* 6, no. 1 (2000): 26–38; R. C. Eaton, R. K. K. Lee, and M. B. Foreman, "The Mauthner Cell and Other Identified Neurons of the Brainstem Escape Response Network," *Progress in Neurobiology* 63 (2001): 467–485. Also, excellent experiments on the variety of ways that fish use their reticulospinal circuits for escape and predation have been conducted by Melina Hale, associate professor of organismal biology and anatomy, University of Chicago.

56. My favorite paper of Fetcho's on the control of swimming behaviors is K. R. Svoboda and J. R. Fetcho, "Interactions Between the Neural Networks for Escape and Swimming in Goldfish," *Journal of Neuroscience* 16, no. 2 (1996): 843–852.

CHAPTER 6

1. From an interview by *New York Times* reporter Adam Bryant, "Don't Lose That Start-Up State of Mind," October 16, 2010, http://www.nytimes.com/2010/10/17/business/17corner.html?_r=1&ref=adam_bryant.

2. I encourage you to explore Full's excellent website: http://polypedal.berkeley.edu/cgi-bin/twiki/view/PolyPEDAL/WebHome.

3. A reference to Ferengi first officer Kazako from "The Battle," an episode of *Star Trek: The Next Generation* (first season).

4. I've tried to address this line of criticism in more detail by placing Tadro and the Tadro evolutionary system in the explicit context of Webb's system: J. H. Long, "Biomimetic Robotics: Building Autonomous, Physical Models to Test Biological Hypotheses," *Proceedings of the Institution of Mechanical Engineers, Part C, Journal of Mechanical Engineering Science* 221 (2007): 1193–1200.

5. Strictly speaking, Rob's observation is a logical inference based on the following reasoning: *post hoc ergo proctor hoc* (after this therefore because of this). In this case, because the three paired sense organs evolved before vertebrae did, then the evolution of vertebrae must be contingent upon those sense organs and the functional capacities that they bring to the vertebrates. We can test this assertion using the new Tadro4 system. Without such a test, by the way, the assertion does not stand on its own accord because many other changes in vertebrates occurred along with the changes in the sense organs.

6. An open-source physics engine that may be used for simulation of rigid-body dynamics is ODE: www.ode.org/. Unfortunately, this physics engine along with most others does not model the interactions of flexible bodies and fluids. That's because the physics are very complicated. Thus, we've been building and modifying our own physics engine.

7. If you are interested in Tom's approach, try: Thomas Ellman, Ryan Deak, and Jason Fotinatos, "Automated Synthesis of Numerical Programs for Simulation of Rigid Mechanical Systems in Physics-Based Animation," *Automated Software Engineering* 10, no. 4 (2003): 367–398.

8. The traits that the engineer allows to vary, just like we allowed traits to vary in our Tadro3, defines the design space. The hill, then, is the combination of those traits that gives the best performance compared to other combinations. Engineers use genetic algorithms instead of doing exhaustive searches of all possible combinations of traits; if you have many traits or features you need to consider, using a genetic algorithm can help you find the hill faster because you don't rely on your intuition to find the optimum (as defined by a position in multiple dimensions, e.g., the optimal design has a specific weight, drag coefficient, and gear ratio).

9. You can find the details of the digi-Tad3 simulation in this paper: J. H. Long Jr., M. E. Porter, C. W. Liew, and R. G. Root, "Go Reconfigure: How Fish Change Shape as They Swim and Evolve," *Integrative and Comparative Biology* 50, no. 6 (2010): 1120–1139.

10. This is a line spoken by Miss Vance, played by Katherine Hepburn in the film *Bringing Up Baby*, as she walked awkwardly following the loss of just one of her high-heeled shoes. I do not mean to imply that Tadro3 limped or wore high heels.

11. Spoken upon the airborne release of an egg trapped in its container by extra-terrestrial Mork, played by Robin Williams, in the television sitcom *Mork and Mindy*.

12. By the way, we are always trying to find ways to automate video analysis. I'll spare you the details, but suffice it to say that we run into three kinds of problems with feature-tracking algorithms: (1) false positives from the light reflected off of the water's surface, (2) abrupt contrast changes as the Tadros pass into and out of regions of high light intensity, and (3) needing to double-check every automatically processed frame for errors. The closest we've come to a nifty solution is to have each Tadro emit an ultrasonic signal that an array of fixed receivers then reads. We simply ran out of money to make this work.

13. For rules, regulations, and results, see www.worldsolarchallenge.org/.

14. Because wobble and speed were correlated, as we saw in Chapter 4, it didn't make sense to use that pair. Also, we did not keep all of the four feeding metrics and then add new predator-avoidance ones on top because we wanted to keep low the number of metrics in order to help us interpret after all was said and done. Finally, average distance from the light is arguably the closest behavioral metric we have to actual light harvesting.

15. This quote is attributed to David Farragut, naval commander during the US Civil War. Although historians question its veracity, it is, no matter its origin, a damn good quote.

16. Interestingly, isolation also works in the same way for evolution. It's on isolated islands, like the Galapagos or the Hawaiian archipelago, that we see rapid evolutionary changes. For a fantastic introduction to rapid evolution on islands and evolutionary processes in general, I recommend this book, mentioned earlier: Jonathan Weiner, *The Beak of the Finch: A Story of Evolution in Our Time* (New York: Alfred A. Knopf, 1994).

17. In fishes the vertebral column is seen to have two main sections, the precaudal and the caudal. The precaudal section is what we haughty mammals might be tempted to call "abdominal" because each vertebra is associated with ribs and an underlying visceral cavity. The caudal section comes after (posterior) the precaudal and, in bony fishes, stops at the caudal fin. In sharks, skates, and rays, however, the vertebral column continues all the way up to the tip of the upper lobe of the caudal fin. Thus, in these cartilaginous fishes it's not clear where the "caudal" vertebrae stop—at the anterior or posterior margin of the caudal fin? Although I'm sure that this is a fascinating topic for most of you, we have to stop. When I say "caudal" here I mean from the last precaudal vertebra to the anterior margin of the caudal fin.

18. Kurt and Keon presented our work on biomimetic vertebral columns at the Annual Meeting of the Society for Integrative and Comparative Biology: K. Bantilan, K. Combie, J. Schaeffer, D. Pringle, J. H. Long Jr., and T. Koob, "Building Biomimetic Backbones: Modeling Axial Skeleton Morphospace," *Integrative and Comparative Biology* 46, suppl. 1 (2006): e8.

19. From the Metallica song, "Enter Sandman" on the album *Metallica*. Rock on!

20. For a detailed comparison of model 3 and model 1 versions, see the following paper: J. H. Long Jr., T. Koob, J. Schaefer, A. Summers, K. Bantilan, S. Grotmol, and M. E. Porter, "Inspired by Sharks: A Biomimetic Skeleton for the Flapping, Propulsive Tail of an Aquatic Robot," *Marine Technology Society Journal* 45, no. 4 (2011): 119–129.

21. The presence of a lateral line is debated, but examination of the different specimens of *Drepanaspis* shows small canals that may have housed neuromast cells. See D. K. Elliot and E. Mark-Kurik, "A Review of the Lateral Line Sensory System in Psammosteid Heterostracans," *Revista Brasileira de Paleontologia* 8, no. 2 (2005): 99–108.

22. C. E. Brett and S. E. Walker, "Predators and Predation in Paleozoic Marine Environments," in *The Fossil Record of Predation,* edited by M. Kowalewski and P. H. Kelley, *Paleontological Society Special Papers* 8 (2002): 93–118.

23. Westneat, an expert on the biomechanics of fish feeding, produced this jaw-dropping analysis: P. S. L. Anderson and M. W. Westneat, "Feeding Mechanics and Bite Force Modeling of the Skull of *Dunkleosteus terrelli*, an Ancient Apex Predator," *Biology Letters* 3, no. 1 (2007): 77–80.

24. An important point to keep in mind here, if you want to try this system in your own Evolvabot, is that because Tadros are surface swimmers, the IR sensors mounted above the water line are working in air.

25. To beat a dead horse, as the saying goes, keep in mind that correlated evolution can turn out to be either concerted or mosaic. In spite of this flogging, you shouldn't be deterred from investigating this character evolution approach. Here's a great place to start: Michael I. Coates and Martin J. Cohn, "Developmental and Evolutionary Perspectives on Major Transformation in Body Organization—Vertebrate Axial and Appendicular Patterning: The Early Development of Paired Appendages," *American Zoologist* 39, no. 3 (1999): 676–685. Also: Robert A. Barton and Paul H. Harvey, "Mosaic Evolution of Brain Structure in Mammals," *Nature* 405, no. 6790 (2000): 1055–1058.

26. D.-G. Shu, S. Conway Morris, J. Han, Z.-F. Zhang, K. Yasui, P. Janvier, L. Chen, X.-L. Zhang, J.-N. Liu, Y. Li, and H.-Q. Lui, "Head and Backbone of the Early Cambrian Vertebrate *Haikouichthys*," *Nature* 421, no. 6922 (2003): 526–529.

27. The term "Kutta condition" describes the physical situation at the trailing edge that is used in the Runge-Kutta theorem of inviscid flow. Runge-Kutta is used to estimate the pattern of flow around a body, including the so-called separation and stagnation points. The pattern of flow around a fish's body is constantly changing as the fish undulates its body, and the caudal fin is the place where water that has interacted with the body is shed rearward, creating a wake. The wake, in turn, can be thought of as evidence of the momentum that the fish has transferred from its body in order to move forward, thanks to Newton's third law.

28. When they started the predator-prey trials for generation six, the students noticed a change in the behavior of the robots that turned out to be related to the servo motors breaking down. The wisdom of their observations had been informed, in part, by the fact that at the start of every generation they ran positive controls on PreyRo and Tadiator, using a fixed "control tail" to assess any degradation in the hardware. Although we normally keep spare and identical parts on hand for just

such a breakdown, we were fresh out of servos and so was the supplier. This pause in evolutionary activities allowed us to analyze this first run, which we published: N. Doorly, K. Irving, G. McArthur, K. Combie, V. Engel, H. Sakhtah, E. Stickles, H. Rosenblum, A. Gutierrez, R. Root, C.-W. Liew, and J. H. Long Jr., "Biomimetic Evolutionary Analysis: Robotically-Simulated Vertebrates in a Predator-Prey Ecology," *Proceedings of the 2009 IEEE Symposium on Artificial Life* (2009): 147–154.

29. Spoken by Don Lockwood, in the movie *Singing in the Rain,* as he attempts to recast his past dowdy Vaudeville career in the color of his current Hollywood persona.

30. I'm playing a bit fast and loose with my use of "concerted" here, so please help me out by keeping in mind that a correlation is only the first piece of evidence for concerted evolution. Because we've defined concerted evolution to be causally based, we also want to show how, in this case, the number of vertebrae improves the acceleration ability of the PreyRo. I'll show you evidence for this in some experiments that we run on the Tadro-derived MARMT system in the next chapter.

CHAPTER 7

1. This opening line is from the movie, *The Adventures of Buckaroo Bonzai Across the Eighth Dimension*, 1984.

2. Even though we spoke at length in Chapter 2 about evolutionary theory, you may hunger for more (or for a refresher). Consider starting at this website, "Understanding Evolution," evolution.berkeley.edu/evolibrary/home.php.

3. You may recognize this problem from our discussion of scientific inference in Chapter 2. We only ever see a limited number of cases of all of the phenomena we seek to understand. Once we infer some property of all possible cases from witnessing just a few, we are always worried, with good reason, about the other cases. What if one of those unseen cases falsifies my idea of how the system is working? With that very real concern in mind, we conduct additional tests, make sure that we sampled the study cases within the system in a way that best represents all of the possible cases, and "prove" by being unable to disprove after repeated attempts to do so.

4. Some folks argue that this question is the motivation for most of evolutionary biology. It was addressed by Sewall Wright, who, in the first part of the twentieth century, extended individual genetics to the genetics of populations. In so doing, he helped propel the modern synthesis of evolutionary theory, which includes his concept of an "adaptive landscape," wherein a population has an evolutionary path that is determined by, you guessed it, history, selection, and random genetic effects. Check out his paper: Sewall Wright, "The Roles of Mutation, Inbreeding, Crossbreeding and Selection in Evolution," *Proceedings of the Sixth International Congress of Genetics* 1 (1932): 356–366. The theory of adaptive landscapes, though modified, is currently used to study, for example, the pathways of molecular evolution: F. J. Poelwijk, D. J. Kiviet, D. M. Weinreich, and S. J. Tans, "Empirical Fitness Landscapes Reveal Accessible Evolutionary Paths," *Nature* 445, no. 7126 (2007): 383–386.

5. This question springs from Steven J. Gould's point about the importance of historical contingency. He argues that chance events in life make it highly unlikely that any species would evolve along the same path given a second opportunity to do so. Steven J. Gould, *Wonderful Life: The Burgess Shale and the Nature of History* (New York: W. W. Norton, 1989). Others argue that chance plays less of a role and that some forms have a high-probability of re-evolving. The genetics of developmental systems may constrain those possibilities. For a great introduction to that subject, read Sean B. Carroll, *Endless Forms Most Beautiful: The New Science of Evo Devo and the Making of the Animal Kingdom* (New York: W. W. Norton, 2005).

6. This question is a variant on these: Why is morphospace clumped? Why is biodiversity limited? What kinds of life-forms are physically possible? Genetically possible?

7. We calculate the selection vector as follows. First, the fitness scores determine who gets to mate. For PreyRo, the top three out of six get to mate, with the first-, second-, and third-place winners contributing to six, four, and two gametes, respectively, to the mating pool. Second, before we mutate those gametes or allow them to join to make offspring, we calculate the average values of the traits of the pre-mutation and pre-mating offspring. Third, this average of the traits is the position of the head of the selection vector's arrow, with the vector's tail anchored at the average of the parental population.

8. Warning: I made the peaks on this map in an intuitive and qualitative manner. In other words, I guessed. Well, it's a bit better than guesswork, but not much, given how little data we've got here. Knowing that the selection vectors point uphill, I knew where at least some peaks or ridges needed to be. The guesswork comes in as follows. For the selection vectors from generations one and two, I assumed that they were pointing to a ridge. I could have assumed that they pointed to two separate peaks. I didn't, though, because the direction that they point is similar, south-south-east for generation one and south-east for generation two, and I took that to mean that they were pointing at the same adaptive structure. This guesswork shows you how much data you would need to create a comprehensive adaptive landscape.

9. C. W. Liew and M. Lahiri, "Exploration or Convergence? Another Meta-Control Mechanism for GAs," in *Proceedings of the 18th International Florida Artificial Intelligence Research Society Conference,* 251–257 (Clearwater Beach, FL: AAAI Press, 2005).

10. "Phenomenal cosmic power! Itty-bitty living space!" said the genie in *Aladdin*, the 1992 Disney film.

11. Thanks to Charles Dickens.

12. General Kurtz in Francis Ford Coppola's 1979 film *Apocalypse Now.*

13. Barbara Webb, "Can Robots Make Good Models of Behaviour?" *Behavioral and Brain Sciences* 24, no. 6 (2001): 1048.

14. Ibid., 1049.

15. Marcel Proust, *À la recherche du temps perdu*, translated by C. K. Scott Moncrieff and Terence Kilmartin as *Remembrance of Things Past* (New York: Vintage Books, 1982).

16. Christopher McGowan, *The Dragon Seekers: How an Extraordinary Circle of Fossilists Discovered the Dinosaurs and Paved the Way for Darwin* (New York: Basic Books, 2001).

17. H. T. de la Beche and W. D. Conybeare, "Notice of the Discovery of a New Animal, Forming a Link Between the *Ichthyosaurus* and Crocodile, Together with General Remarks on the Osteology of *Ichthyosaurus*," *Transactions Geological Society London* 5 (1821): 559–594.

18. Richard Forrest has created and maintains an excellent site on plesiosaurs that you should visit: plesiosaur.com/. Dr. Adam Stuart Smith also has an excellent site that features his own research: www.plesiosauria.com/index.html.

19. The term "plesiosaur" can be confusing. For example, within the Order Plesiosauria, we've got the short-necked Pliosauroidea and the long-necked Plesiosauroidea as Suborders. When I use the term plesiosaur here, I include all members of the Order, after Adam Stuart Smith (www.plesiosauria.com/classification.html). Just keep in mind that some folks prefer to talk about "true plesiosaurs" as just the long-necked forms, leaving pliosaurs to the side.

20. Carl Zimmer, *At the Water's Edge: Fish with Fingers, Whales with Legs, and How Life Came Ashore but Then Went Back to Sea* (New York: Touchstone, 1998).

21. J. Lindgren, M. W. Caldwell, T. Konishi, and L. M. Chiappe, "Convergent Evolution in Aquatic Tetrapods: Insights from an Exceptional Fossil Mosasaur," *PLoS One* 5, no. 8 (2010): e11998, doi:10.1371/journal.pone.0011998.

22. Start here with two of Frank's papers: "Transitions from Drag-Based to Lift-Based Propulsion in Mammalian Aquatic Swimming," *American Zoologist* 36, no. 5 (1996): 628–641, and "Biomechanical Perspective on the Origin of Cetacean Flukes," in *The Emergence of Whales: Evolutionary Patterns in the Origin of Cetacea*, edited by J. G. M. Thewissen, 303–324 (New York: Plenum Press, 1998).

23. These flippers, or Nektors, are themselves biologically inspired. Charles Pell, working with a graduate student at Duke University in the BioDesign Studio that he and Professor Steve Wainwright created, noticed that a fish-like piece of rubber, mounted on a stick, would generate thrust if you wiggled the stick between your fingers, rolling the stick between thumb and forefinger, with the fish in the water. Pell, then-student-of-mine Matt McHenry, and I used Nektors as model representations of blue-gill sunfish to analyze swimming propulsion: M. J. McHenry, C. A. Pell, and J. H. Long Jr., "Mechanical Control of Swimming Speed: Stiffness and Axial Wave Form in an Undulatory Fish Model," *Journal of Experimental Biology* 198 (1995): 2293–2305. Pell and Wainwright patented the Nektor system: C. A. Pell, and S. A. Wainwright, "Swimming Aquatic Creature Simulator," US Patent 6179683, issued January 30, 2001, assigned to Nekton Technologies, Inc. (now the marine division of iRobot, Inc.).

24. Tellingly, petit madeleines are modeled after scallops! If you buy madeleine pans, you'll notice right away the fluted and streamlined depressions into which you pour the batter. What's cool about scallops is that they are bivalves, mollusks with two shells, that actually swim. So here we have a swimming scallop that is the model for a streamlined pastry that is the inspiration of the name of a swimming and streamlined biorobot. Does it get any more fun?

25. Forgive the engineer-speak about to issue forth. Maddie's flippers, or Nektors, are single-degree-of-freedom actuators. A shaft colinear with a rotary motor moves the flipper, the compliant material forming the shape and bulk of the appendage, which is molded around the shaft at a specified angular velocity. The flipper is oriented so that its leading edge rotates in pitch. That pitch rotation flaps the flipper and transfers angular momentum to the surrounding water. When the pitch rotation is reciprocated such that the direction of the angular velocity alters regularly, as with a sine function, then the momentum transferred from the flipper to the water can be focused as a jet. This jet, in turn, produces a net thrust, via Newton's third law, on the oscillating flipper.

26. The coaches of Vassar's swim teams, Lisl Prater-Lee, Tom Albright, and Jesup Szatkowski, were kind enough to allow Robot Madeleine both training and experiment time in the pool.

27. You can find all of the details of this set of experiments in the following paper: J. H. Long Jr., J. Schumacher, N. Livingston, and M. Kemp, "Four Flippers or Two? Tetrapodal Swimming with an Aquatic Robot," *Bioinspiration & Biomimetics* (Institute of Physics) 1 (2006): 20–29. We first introduced Robot Madeleine here: M. Kemp, B. Hobson, and J. H. Long Jr., "Madeleine: An Agile AUV Propelled by Flexible Fins," in *Proceedings of the 14th International Symposium on Unmanned Untethered Submersible Technology (UUST)*, Autonomous Undersea Systems Institute, Lee, NH, 2005.

28. F. E. Fish, J. Hurle, and D. P. Costa, "Maneuverability by the Sea Lion *Zalophus californianus*: Turning Performance of an Unstable Body Design," *The Journal of Experimental Biology* 206, pt. 4 (February 2003): 667–674.

29. Predator X is the stage name of a heretofore undescribed species of pliosaur unearthed in the Norwegian Arctic. The History Channel aired an eponymous special on Predator X, and clips of the documentary are available at www.history.com/videos/predator-x-revealed#predator-x-revealed. Robot Madeleine, by the way, was featured!

30. The accelerations of twenty-two-meter- to twenty-seven-meter-long blue whales have been measured in the wild: J. A. Goldbogen, J. Calambodkidis, E. Oleson, J. Potvin, N. D. Pyenson, G. Schorr, and R. E. Shadwick, "Mechanics, Hydrodynamics and Energetics of Blue Whale Lunge Feeding: Efficiency Dependence on Krill Density," *The Journal of Experimental Biology* 214, no. 1 (2011): 131–146.

31. Robot Madeleine, like Tadro, has had multiple versions. Maddie 1.0 was self-propelled and controlled remotely by a human operator. Maddie 2.0 had all the on-board sensors, like the power monitor and the accelerometer, allowing her to collect data on herself. Maddie 2.0 was the version that I've talked about here and about which we've published our papers. Maddie 3.0 was programmed by Mathieu Kemp to be fully autonomous, employing a two-layer subsumption hierarchy (see Chapter 5) in which she selected a random depth and compass heading and then moved along that course until she either detected an object with her sonar or ran out of time (thirty seconds). Maddie 3.0 was destroyed, unfortunately, when we were filming her for the documentary *Predator X*; she sprung a leak and fried her

electronics. Since then we have been trying to rebuild her as Maddie 4.0 at Vassar; however, at the moment we lack the funding to finish that job.

32. You can see Transphibians at the iRobot website: www.irobot.com/gi/maritime/Transphibian/.

33. B. W. Hobson, M. Kemp, R. Moody, C. A. Pell, and F. Vosburgh, "Amphibious Robot Devices and Related Methods," US Patent 6,974,356, 2005.

34. Broadcast date of August 9, 2006.

35. Auke Jan Ijspeert, Alessandro Crespi, Dimitri Ryczko, and Jean-Marie Cabelguen, "From Swimming to Walking: Is a Salamander Robot Driven by a Spinal Cord Model?" *Science* 315, no. 5817 (2007): 1416–1420.

36. You can read more about MARMT in J. H. Long Jr., N. Krenitsky, S. Roberts, J. Hirokawa, J. de Leeuw, and M. E. Porter, "Testing Biomimetic Structures in Bioinspired Robots: How Vertebrae Control the Stiffness of the Body and the Behavior of Fish-like Swimmers," *Integrative and Comparative Biology* 51, no. 1 (2011): 158–175, doi:10.1093/icb/icr020.

CHAPTER 8

1. Where would we be without Douglas Adams? This chapter title is an homage to the fourth book in his *Hitchhiker's Guide to the Galaxy* series, *So Long, and Thanks for All the Fish* (New York: Harmony Books, 1985).

2. Here's MHI's original press release: www.mhi.co.jp/en/news/sec1/e_0898.html.

3. You can learn more about this company's plans at www.robotswim.com.

4. Reported by the *Huffington Post*, July 16, 2010, based on a Reuters video posted July 15, 2010. Or, better yet, visit Dr. Porfiri's web page for the real scoop: faculty.poly.edu/~mporfiri/index.htm.

5. Full disclosure here: I have been and currently am collaborating with Farshad and FarCo Technologies. However, I hold no financial stake in FarCo Technologies (www.farcotech.com/).

6. P. R. Bandyopadhyay, "Swimming and Flying in Nature—The Route Toward Applications: The Freeman Scholar Lecture," *Journal of Fluids Engineering* 131, no. 3 (March 2009): 0318011–0318029.

7. Steven Vogel, *Cats' Paws and Catapults: Mechanical Worlds of Nature and People* (New York: W. W. Norton, 1998), 10.

8. E-mail message from Melina Hale, January 7, 2011.

9. Full disclosure: I am hired by the European Commission as an outside expert evaluator of the FILOSE project, which the EC funds as part of their Seventh Framework Programme.

10. For the latest on FILOSE Fish, see this paper: M. Kruusmaa, T. Salumae, G. Toming, A. Ernits, and J. Ježov, "Swimming Speed Control and On-board Flow Sensing of an Artificial Trout," *Proceedings of the IEEE International Conference of Robotics and Automation* (IEEE ICRA 2011), Shanghai, China, May 9–13, 2011.

11. See the full statement at this URL: cordis.europa.eu/fp7/understand_en.html.

12. Work on the fish and the biomimetic robot is explained in: O. M. Curet, N. A. Patankar, G. V. Lauder, and M. A. MacIver, "Aquatic Manoeuvering with Counter-Propagating Waves: A Novel Locomotive Strategy," *Journal of the Royal Society Interface* 8, no. 60 (July 2011), 1041–1050, doi:10.1098/rsif.2010.0493.

13. For more on their robotic fish fin, see: Chris Phelan, James Tangorra, George Lauder, and Melina Hale, "A Biorobotic Model of the Sunfish Pectoral Fin for Investigations of Fin Sensorimotor Control," *Bioinspiration & Biomimetics* 5, no. 3 (2010); James Louis Tangorra, S. Naomi Davidson, Ian W. Hunter, Peter G. A. Madden, George V. Lauder, Dong Haibo, Meliha Bozkurttas, and Rajat Mittal, "The Development of a Biologically Inspired Propulsor for Unmanned Underwater Vehicles," *IEEE Journal of Oceanic Engineering* 32, no. 3 (2007): 533–550.

14. If you are interested in other fish-inspired robots, I review the field in "Biomimetics: Robotics Based on Fish Swimming," in *Encyclopedia of Fish Physiology: From Genome to Environment*, vol. 1, edited by A. P. Farrell, 603–612 (San Diego: Academic Press, 2011).

15. Conversation at the Annual Meeting of the Society for Integrative and Comparative Biology, January 4, 2011.

16. I use the date of 1946 here because that was when President Truman created the ONR to "plan, foster and encourage scientific research in recognition of its paramount importance as related to the maintenance of future naval power, and the preservation of national security." The Navy, however, considers ONR to have been started earlier, in 1923, as the Naval Research Laboratory. See their timeline at www.onr.navy.mil/About-ONR/History-ONR-Timeline.aspx.

17. In vibratory mechanics the natural frequency of a structure is proportional to the square root of its stiffness. Other factors, like mass and damping, shouldn't be neglected because they play huge roles in how the structure moves.

18. Details of the experiments that originally led us to this prediction can be found in the following paper: J. H. Long Jr., M. J. McHenry, and N. C. Boetticher, "Undulatory Swimming: How Traveling Waves Are Produced and Modulated in Sunfish (*Lepomis gibbosus*)," *Journal of Experimental Biology* 192 (1994): 129–145.

19. You can read more about these early robotic fish in the following paper: M. J. McHenry, C. A. Pell, and J. H. Long Jr. "Mechanical Control of Swimming Speed: Stiffness and Axial Wave Form in an Undulatory Fish Model," *Journal of Experimental Biology* 198 (1995): 2293–2305.

20. For a summary of the ONR's biorobotics program through 2005, see P. R. Bandyopadhya, "Trends in Biorobotic Autonomous Undersea Vehicles," *IEEE Journal of Oceanic Engineering* 30, no. 1 (2005): 109–139.

21. DARPA's mission statement can be found at www.darpa.mil/mission.html.

22. For more on the design and performance of RiSE, see M. J. Spenko, G. C. Haynes, J. A. Saunders, M. R. Cutkosky, A. A. Rizzi, R. J. Full, and D. E. Koditschek, "Biologically Inspired Climbing with a Hexapedal Robot," *Journal of Field Robotics* 25, no. 4 (2008): 223–242.

23. For example, see DARPA CBS-ONR-ARL US Navy Marine Mammal Program, Biosonar Program Office, SPAWAR Systems Center, San Diego, CA, 2002.

See also Frank E. Fish, "Review of Natural Underwater Modes of Propulsion," DARPA, 2000. More recent projects include bio-inspired underwater sensing and autonomous underwater navigation in rivers and estuaries. For more on the workings of DARPA, I recommend Michael Belfiore, *The Department of Mad Scientists: How DARPA Is Remaking our World, from the Internet to Artificial Limbs* (Washington, DC: Smithsonian Books, 2009).

24. I checked DARPA's public solicitation on January 8, 2011, at www.darpa.mil/openclosedsolicitations.html.

25. As reported by John Markoff, "War Machines: Recruiting Robots for Combat," *New York Times*, November 27, 2010.

26. Professor Arkin's book is timely and opens up an important discussion: Ronald C. Arkin, *Governing Lethal Behavior in Autonomous Robots* (Boca Raton, FL: Chapman & Hall/CRC, 2009).

27. Nowadays, the US Coast Guard has eleven missions: www.uscg.mil/top/missions/.

28. This is the translation given by Gilbert in his comprehensive book: Martin Gilbert, *The First World War: A Complete History* (New York: Henry Holt, 1994), 352. Horace's phrase has other translations, including, "It is sweet and right to die for your country."

29. This is an excerpt of Owen's "Dulce et Decorum Est," which can be found in full and with notes at the War Poetry website: www.warpoetry.co.uk/owen1.html.

30. Michael Herr, *Dispatches* (New York: Alfred A. Knopf, 1977).

31. Peter and Craig's model can be found in this article: P. J. Czuwala, C. Blanchette, S. Varga, R. G. Root, and J. H. Long Jr., "A Mechanical Model for the Rapid Body Flexures of Fast-Starting Fish," in *Proceedings of the 11th International Symposium on Unmanned Untethered Submersible Technology (UUST)*, 415–426 (Lee, NH: Autonomous Undersea Systems Institute, 1999). At the same meeting Rob presented this paper: R. G. Root, H-W. Courtland, C. A. Pell, B. Hobson, E. J. Twohig, R. J. Suter, W. R. Shepherd, III, N. Boetticher, and J. H. Long Jr., "Swimming Fish and Fish-like Models: The Harmonic Structure of Undulatory Waves Suggests That Fish Actively Tune Their Bodies," in *Proceedings of the 11th International Symposium on Unmanned Untethered Submersible Technology (UUST)*, 378–388 (Lee, NH: Autonomous Undersea Systems Institute, 1999).

32. The irony is that secrecy is enforced when I work with and advise companies. Both business and the military use secrecy to maintain an advantage over the competition or adversaries. For the record, I honor all of my agreements with businesses to keep our proprietary work secret.

33. The race continues unabated: E. Bumiller and T. Shanker, "War Evolves with Drones, Some Tiny as Bugs," *New York Times*, June 19, 2011.

34. For the latest on autonomous robots in war: L. G. Weiss, "Autonomous Robots in the Fog of War," *IEEE Spectrum* 48, no. 8 (2011): 30–57.

35. Silke Steingrube, Marc Timme, Florentin Worgotter, and Poramate Manoonpong, "Self-Organized Adaptation of a Simple Neural Circuit Enables Complex Robot Behaviour," *Nature Physics* 6, no. 3 (2010): 224–230.

36. Two groundbreaking papers by Lipson that you simply must read: H. Lipson and J. B. Pollack, "Automatic Design and Manufacture of Artificial Lifeforms," *Nature* 406, no. 6799 (2000): 974–978; and J. Bongard, V. Zykov, and H. Lipson, "Resilient Machines Through Continuous Self-Modeling," *Science* 314, no. 5802 (November 2006): 1118–1121.

37. Bongard explains his approach on his web page: www.cs.uvm.edu/~jbongard/research.html.

38. Penrose the Elder summarized their work in this article: L. S. Penrose, "Self-Reproducing Machines," *Scientific American* 200, no. 6 (June 1959): 105–114.

39. You can watch the Penrose machines replicating in this film, made in 1961: http://vimeo.com/10298933.

40. MicroHunter was invented by Chuck Pell, Hugh Crenshaw, Jason Janet, and Mathieu Kemp and was assigned to Nekton Technologies, Inc., US Patent 6,378,801. C. Pell, H. Crenshaw, J. Janet, and M. Kemp, "Devices and Methods for Orienting and Steering in Three-Dimensional Space," 2002. A great place to get an overview of MicroHunter is in J. Wakefield, "Mimicking Mother Nature," *Scientific American* 286, no. 1 (January 2002): 26–27.

41. For more on MicroHunter, see M. Kemp, H. Crenshaw, B. Hobson, J. Janet, R. Moody, C. Pell, H. Pinnix, and B. Schulz, "Micro-AUVs I: Platform Design and Multiagent System Development," in *Proceedings of the 12th International Symposium on Unmanned Untethered Submersible Technology (UUST)*, 2001.

42. For more on Navy SEALs, see their website: www.navyseal.com/navy_seal/.

43. US Army Field Manual (FM) 100–105, *Operations* (Washington, DC: Government Printing Office [GPO], 1993), 6.

44. R. H. Kewley and M. J. Embrechts, "Computational Military Tactical Planning System," *IEEE Transactions on Systems, Man, and Cybernetics, Part C: Applications and Reviews* 32, no. 2 (2002): 161–171.

45. L. G. Shattuck, "Communicating Intent and Imparting Presence," *Military Review* 80, pt. 2 (March–April 2000): 66–72.

46. Ronald Arkin, an expert robotics engineer, is the leader in considering both the practical and philosophical aspects of the ethics of using robots in war. His first paper on the subject is a good place to start: "Governing Lethal Behavior: Embedding Ethics in a Hybrid Deliberative/Reactive Robot Architecture—Part 1: Motivation and Philosophy," *Proceedings of Human-Robot Interaction 2008*, Amsterdam, Netherlands, 2008.

47. Ronald Arkin, *Governing Lethal Behavior in Autonomous Robots* (Boca Raton, FL: Chapman & Hall/CRC, 2009), 2.

INDEX